# Management System for Safety

## Jeremy Stranks

PITMAN
PUBLISHING

PITMAN PUBLISHING
128 Long Acre, London WC2 9AN

A Division of Longman Group UK Limited

First published in Great Britain, 1994

**British Library Cataloguing in Publication Data**
A CIP catalogue record for this book can be obtained from the British Library.

ISBN 0 273 60441 4

Typeset by Northern Phototypesetting Co Ltd., Bolton
Printed and bound in Great Britain by
Bell & Bain Ltd., Glasgow

*The Publishers' policy is to use paper manufactured from sustainable forests*

# Contents

# Introduction

The Management of Health and Safety at Work Regulations 1992 came into operation on 1 January 1993. Apart from the absolute or strict nature of many of the new duties laid on employers under these Regulations, they bring in a new concept, namely that of charging employers with the duty to *actively manage* those activities aimed at protecting the health and safety of people at work.

Previous legal requirements, such as those under the Factories Act 1961 and Offices, Shops and Railway Premises Act 1963 laid down prescriptive or minimum standards on, for instance, the minimum temperature in a workplace, the total number of wash hand basins required according to the number of employees, or the specific safety features of a scaffold. There was no clear-cut duty to actually manage their health and safety activities, however. Moreover, many of these pre-1993 duties are qualified by the term 'so far as is reasonably practicable', implying a balance between the cost of sacrifice involved on the one hand and the risk to people on the other. On this basis, an employer did not necessarily have to comply if it could be proved to a court that compliance was not reasonably practicable.

Whilst much of this legislation is still in force, it is clear that the legal trend is now away from the compliance with minimum legal requirements to that of placing the duty on managers to introduce management systems, implement these systems and monitor the effectiveness of same. Clearly, many organisations need to re-examine their health and safety strategies and look to the development of the right safety culture, one based on commitment, leadership, management example, training, ownership at all levels, the setting of realistic and achievable targets and adequate resourcing.

This book examines the principles and objectives of health and safety management. It also provides guidance on the practical implementation of the requirements of the Management of Health and Safety at Work Regulations, such as the procedures for risk assessment, the designation and training of competent persons and techniques for monitoring health and safety performance. Areas such as occupational health management, an aspect sadly neglected in many organisations, and joint consultation are also covered, along with safety management techniques, such as loss control and risk management.

Jeremy Stranks, 1994

# List of abbreviations

| | |
|---|---|
| ACOP | Approved Code of Practice |
| CBI | Confederation of British Industries |
| COSHHR | Control of Substances Hazardous to Health Regulations 1988 |
| EMAS | Employment Medical Advisory Service |
| ETBA | Energy Trace and Barrier Analysis |
| FA | Factories Act 1961 |
| HAZOPS | Hazard and Operability Studies |
| HSE | Health and Safety Executive |
| HSC | Health and Safety Commission |
| HSWA | Health and Safety at Work etc. Act 1974 |
| MORT | Management Oversight and Risk Tree |
| MPL | Maximum possible loss |
| MHOR | Manual Handling Operations Regulations 1992 |
| MHSWR | Management of Health and Safety at Work Regulations 1992 |
| OSRPA | Offices, Shops and Railway Premises Act 1963 |
| PPE | Personal protective equipment |
| PPEWR | Personal Protective Equipment at Work Regulations 1992 |
| PUWER | Provision and Use of Work Equipment Regulations 1992 |
| RIDDOR | Reporting of Injuries, Diseases and Dangers Occurrences Regulations 1985 |
| RoSPA | Royal Society for the Prevention of Accidents |
| SRSCR | Safety Representatives and Safety Committees Regulations 1977 |
| THERP | Technique for Human Error Rate Probability |
| TLC | Total Loss Control |
| WHSWR | Workplace (Health, Safety and Welfare) Regulations 1992 |

# Principles and objectives of health and safety management

## MANAGEMENT PRINCIPLES

What is management? One definition is:

'The effective use of resources in the pursuit of organisational goals.'

'Effective' implies achieving a balance between the risk of being in business and the cost of eliminating or reducing those risks. Management entails leadership, authority and co-ordination of resources, together with:

(a)  planning and organisation;

(b)  communication;

(c)  selection and training of subordinates;

(d)  accountability and responsibility.

### Health and safety management

Health and safety management is no different from other forms of management. It covers:

(a)  the management of the health and safety operation at national and local level – setting of policy and objectives, organising, controlling, and establishing accountability

(b)  measurement of health and safety performance on the part of the individuals and specific locations

(c)  motivating managers to improve standards of health and safety performance in those areas under their control.

Decision-making is a very important feature of the management process. It can be summarised thus:

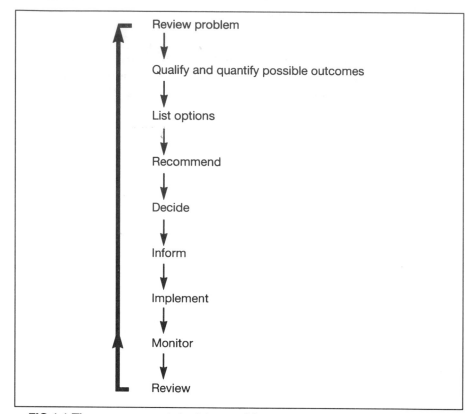

• **FIG 1.1 The management decision-making process**

Management is concerned with people at all levels of the organisation and human behaviour, in particular personal factors such as attitude, perception, motivation, personality, learning and training. Communication brings these various behavioural factors together.

The management of health and safety is also concerned with organisational structures, the climate for change within an organisation, individual roles within the organisation and the problem of stress, which takes many forms. Several questions must be asked at this stage.

1. Are managers good at managing health and safety?
2. What is their attitude to health and safety? Is it of a pro-active or reactive nature?
3. How do their attitudes affect the development of health and safety systems?
4. Is health and safety a management tool, or a problem thrust on them by enforcement agencies?

## *Management approaches to health and safety*

Approaches to health and safety vary considerably at all levels of management. They can be categorised thus:

### The legalistic approach

This classic approach says 'Comply with the law, but no more!' In other words, the organisation is prepared to do the minimum to keep out of the clutches of the enforcement agencies.

### The socio-humanitarian approach

This approach considers the human resources aspects, namely, that people are an important feature of the business operation and an important asset. As such, they must be protected.

### The financial–economic approach

The maintenance of high standards of health and safety needs adequate financial resources. However, all accidents, incidents and occupational ill-health cost money. Most organisations are good at calculating the costs of health and safety improvement, e.g. machinery guarding, health and safety training, but are not good at calculating the losses associated with accidents, sickness, property damage incidents and generally poor health and safety performance. These losses tend to get absorbed in the operating costs of the business. Accident and sickness costing systems soon identify the costs to the organisation of poor health and safety performance, particularly when such identified costs are drawn against individual manager's budgets. This can be the most significant motivator for bringing about improvement.

### The human factors approach

With the greater emphasis on the human factors approach to health and safety, there is a need to actually identify those organisational characteristics that influence safety-related behaviour. These include:

(a) the need to produce a positive climate in which health and safety is seen by both management and employees as being fundamental to the organisation's day-to-day operations, i.e. a positive, safety culture

(b) the need to ensure that policies and systems which are devised for the control of risk from the organisation's operations take proper account of human capabilities and fallibilities

(c) commitment to the achievement of progressively higher standards which is shown at the top of the organisation and down through successive levels of it

(d) demonstration by senior management of their active involvement, thereby stimulating managers throughout the organisation into action

(e) leadership, whereby an environment is created which encourages safe behaviour.

## Identifying the priorities

Whilst health and safety management covers many areas, there are a number of aspects which are significant, namely:

- the company statement of health and safety policy;
- procedures for health and safety monitoring and performance measurement;
- clear identification of the objectives and standards which must be measurable and achievable by the persons concerned;
- the system for improving knowledge, attitudes and motivation and for increasing individual awareness of health and safety issues, responsibilities and accountabilities;
- procedures for eliminating potential hazards from plant, machinery, substances and working practices through the design and operation of safe systems of work and other forms of hazard control;
- measures taken by management to ensure legal compliance.

# PRO-ACTIVE AND REACTIVE MANAGEMENT STRATEGIES

## Pro-active strategies

Any management system should be of a mainly pro-active nature, concerned with the prevention of accidents and ill-health, control of the working environment and the various factors affecting that environment. However, there will always be a place for the various reactive strategies as a means of obtaining feedback on performance.

Generally, accidents are unforeseeable, as far as the accident vicitim is concerned, unplanned, unintended and unexpected. In most cases there is a sequence of events leading to the accident and there may be a number of direct and indirect contributory causes. Industrial accidents can result in death and personal injury; ill-health; damage to property, plant and services; and stoppage of, or interference with, work processes. Fundamentally, they interrupt the business operation and represent some level of loss to the organisation. Pro-active strategies are, therefore, concerned with prevention and can be classified as either 'safe place' or 'safe person' strategies.

### Safe place strategies

The principal objective of a 'safe place' strategy is that of bringing about a reduction in the objective danger to people at work. These strategies feature in much of the occupational health and safety legislation that has been enacted over the last century, such as the Factories Acts 1937 and 1961 (FA); the Offices, Shops and Railway Premises Act 1963 (OSRPA); the Mines and Quarries Act 1954; and, more recently, the Health and Safety at Work etc. Act

1974 (HSWA) in terms of, for instance, basic requirements relating to machinery safety, electrical safety and construction safety. The Workplace (Health, Safety and Welfare) Regulations 1992 (WHSWR) and the Provision and Use of Work Equipment Regulations 1992 (PUWER) continue this philosophy. 'Safe place' strategies may be classified as follows:

### 1. Safe premises
This relates to the general structural requirements of workplaces, such as the stability of buildings, soundness of floors and the load-bearing capacity of beams. Environmental working conditions, such as the levels of lighting, ventilation and humidity, feature in this classification.

### 2. Safe plant
A wide range of plant and machinery and other forms of work equipment, their power sources, location and use are relevant in this case. The safety aspects of individual processes, procedures for vetting and testing new machinery and plant, and systems for maintenance and cleaning must be considered.

### 3. Safe processes
All factors contributing to the operation of a specific process must be considered, e.g. work equipment, raw materials, procedures for loading and unloading, the ergonomic aspects of machine operation, hazardous substances used, and the operation of internal factory transport such as fork-lift trucks.

### 4. Safe materials
Significant in this case are the health and safety aspects of potentially hazardous chemical substances, radioactive substances, raw materials of all types and specific hazards associated with the handling of these materials. Adequate and suitable information on their correct use, storage and disposal must be provided by manufacturers and suppliers, and there may be a need to assess potential health risks in accordance with the Control of Substances Hazardous to Health Regulations 1988 (COSHH).

### 5. Safe systems of work
A safe system of work is defined as 'the integration of people, machinery and materials in a correct working environment to provide the safest possible working conditions'. The design and implementation of safe systems of work is, perhaps, the most important 'safe place' strategy. It incorporates planning, involvement of operators, training, and designing out hazards which may have existed with previous systems.

### 6. Safe access to and egress from work
This refers to access to, and egress from, both the workplace from the road outside and the working position, which may be some distance from the

ground, as with construction workers, or several miles below ground in the case of miners. Consideration must be given, therefore, to workplace approach roads, yards, work at high level, the use of portable and fixed access equipment (e.g. ladders and lifts) and the shoring of underground workings.

### 7. Adequate supervision

HSWA requires that in all organisations there must be adequate safety supervision directed by senior management through supervisory management to workers.

### 8. Competent and trained personnel

The need to appoint competent persons is laid down in the Construction Regulations, other specific Regulations and, more recently, the Management of Health and Safety at Work Regulations 1992 (MHSWR). The general duty to train staff and others is laid down in HSWA, sec 2(2)(c). This duty is further reinforced and extended and specified under the MHSWR. All employees need some form of health and safety training. This should be undertaken through induction training and on being exposed to new risks, on a change of responsibilities, the introduction of new equipment, the introduction of new technology or new systems of work. A well-trained labour force is a safe labour force and in organisations which carry out health and safety training accident and ill-health costs tend to be lower.

### 'Safe person' strategies

Generally, 'safe place' strategies provide better protection than 'safe person' strategies. However, where it may not be possible to operate a 'safe place' strategy, then a 'safe person' strategy must be used. In certain cases, a combination of both strategies may be appropriate.

The main aim of a 'safe person' strategy is to increase people's perception of risk. One of the principal problems of such strategies is that they depend upon the individual conforming to certain prescribed standards and practices, such as the use of certain items of personal protective equipment. Control of the risk is, therefore, placed in the hands of the person whose appreciation of the risk may be lacking or even non-existent. 'Safe person' strategies may be classified as follows:

### 1. Care of the vulnerable

In any work situation there will be some people who are more vulnerable to certain risks than others, e.g. where such workers may be exposed to toxic substances, to small levels of radiation, or to dangerous metals, such as lead. Typical examples of 'vulnerable' groups are young persons who, through their lack of experience, may be unaware of hazards; pregnant women, where there may be a specific risk to the unborn child; and disabled persons, whose capacity to undertake certain tasks may be limited. In a number of cases there may be a need for continuing medical and/or health surveillance of such persons.

## 2. Personal hygiene

The risk of occupational skin conditions caused by contact with hazardous substances such as solvents, glues, adhesives and a wide range of chemical skin sensitisers, needs consideration. There may also be the risk of ingestion of hazardous substances as a result of contamination of food and drink and their containers. In order to promote good standards of personal hygiene it is vital that the organisation provides adequate washing facilities – wash basins, showers and drying facilities – for use by workers, particularly prior to the consumption of food and drink and to returning home at the end of the work period.

## 3. Personal protective equipment

Generally, the provision and use of any items of personal protective equipment must be seen either as a last resort when all other methods of protection have failed or an interim method of protection until some form of 'safe place' strategy can be put into operation. It is by no means a perfect form of protection in that it requires the person at risk to use or wear the equipment all the time they are exposed to a particular hazard. People simply will not always do this!

The Personal Protective Equipment at Work Regulations 1992 (PPEWR) define 'personal protective equipment' (PPE) as meaning 'all equipment (including clothing affording protection against the weather) which is intended to be worn or held by a person at work and which protects him against one or more risks to his health or safety, and any addition or accessory designed to meet that objective'. PPE can be classified thus:

(a) *Head, face and neck protection* – safety helmets, caps with snood attachment, face shields

(b) *Eye protection* – safety spectacles, goggles and face shields

(c) *Ear Protection* – ear defenders, muffs and pads, ear plugs, ear valves

(d) *Respiratory protection* – face masks, general purpose dust respirators, positive pressure-powered dust respirators, helmet-contained pressure respirators, gas respirators, emergency escape respirators, airline breathing apparatus, self-contained breathing apparatus

(e) *Skin protection* – barrier creams

(f) *Body protection* one and two-piece overalls, aprons, donkey jackets, warehouse coats

(g) *Hand and arm protection* – general purpose fibre gloves, PVC fabric gauntlets and sleeves, chain mail gloves and arm protectors

(h) *Leg and foot protection* – safety boots, wellingtons and shoes, gaiters.

The principal effect of the PPEWR is to make it far more difficult for employers to rely on the provision of PPE as the sole means of protecting people at work. Under these Regulations employers must ensure that, where more than one risk to health or safety makes it necessary for an employee to wear or use more than one item of PPE, such equipment is compatible and continues to be effective against the risk or risks in question (Reg. 5). Before choosing any PPE that he is required to provide, an employer must ensure an assessment is made to determine whether what he intends to provide is suitable (Reg. 6). Furthermore, PPE provided to employees must be maintained (including replaced or cleaned as appropriate) in an efficient state, in efficient working order and good repair (Reg. 7). Accommodation must also be provided for the storage of PPE when it is not being used (Reg. 8). Employees required to wear or use PPE must also be provided with such information, instruction and training as is adequate and appropriate to enable the employee to know:

(a)  the risk or risks which the PPE will avoid or limit

(b)  the purpose for which and manner in which the PPE is to be used

(c)  any action to be taken by the employee to ensure that the PPE remains in an efficient state, in efficient working order and in good repair (Reg. 9)

Under Reg. 10 employers must take all reasonable steps to ensure that any PPE is properly used. Employees must use any PPE provided in accordance both with any training received and the instructions respecting that use which have been provided. Employees must take all reasonable steps to ensure that the PPE is returned to the accommodation provided for it after use. Reg. 11 requires employees to report any loss of or obvious defect in the PPE to their employer.

### 4. Safe behaviour
Employees must not be allowed to indulge in unsafe behaviour or 'horseplay'. Examples of unsafe behaviour include the removal, or defeating the purpose of, machinery guards and safety devices, smoking in designated 'no smoking' areas, the dangerous driving of vehicles and the failure to wear or use certain items of PPE.

### 5. Caution towards danger
All workers and management should appreciate the risks in the workplace, and these risks should be clearly identified in the Statement of Health and Safety Policy required under HSWA, together with the precautions required to be taken by workers to protect themselves from such risks.

## Reactive strategies

Whilst the principal effort must go into the implementation of pro-active strategies, it is generally accepted that there will always be a need for reactive or 'post-accident' strategies, particularly as a result of failure of the various 'safe person' strategies. The problem with people is that they forget, they take short cuts to save time and effort, they sometimes do not pay attention or they may consider themselves too experienced and skilled to bother about taking basic precautions. Reactive strategies can be classified as follows:

### 1. Disaster/contingency/emergency/planning
Here there is a need for managers to ask themselves this question: what is the very worst possible type of incident or event that could arise in the business activity? For most types of organisation this could be a major escalating fire, but other types of major incident should be considered, such as an explosion, collapse of a scaffold, flood or serious vehicular traffic accident. The need for some form of emergency plan should be considered. See also Chapter 4 – *Safety Management Systems*.

### 2. Feedback strategies
Accident and ill-health reporting, recording and investigation provide feedback as to the indirect and direct causes of accidents. The study of past accident causes provides information for the development of future pro-active strategies. The limitations of accident data as a measure of safety performance are covered in Chapter 4 – *Safety Management Systems*.

### 3. Improvement/ameliorative strategies
These strategies are concerned with minimising the effects of injuries as quickly as possible following an accident. They will include the provision and maintenance of first-aid services, procedures for the rapid hospitalisation of injured persons, and possibly, a scheme for rehabilitation following major injury.

## INSURANCE COMPANIES AND RISK MANAGEMENT

Insurance companies over the last thirty years have provided organisations with a new concept in dealing with the risks associated with their business and other forms of activity. 'Risk management' can be defined as 'the minimisation of the adverse effects of pure risks, namely those risks that only result in a loss to the organisation'. The technique fundamentally involves the identification, evaluation and economic control of the risks to which a business is exposed. Apart from examining risks to the health and safety of staff and the public, risk management services consider aspects like damage control, the risks associated with the operation of vehicle fleets, fire, security, pollution, product liability, public liability and computer-related risks, all of which can result in significant losses to organisations.

'Risk management' is dealt with in greater detail in Chapter 7 – *Risk Management*.

## COMMUNICATION SYSTEMS

Communication is the transfer of information, ideas and emotions between one individual or group of individuals and another. The objectives of communication are to understand others, to obtain clear perception or reception of information, to obtain understanding, to achieve acceptance (i.e. agreement and commitment, of ideas), and to facilitate or obtain effective human behaviour or action.

Sound communication on health and safety-related issues is a vital feature of effective health and safety management. Whilst the subject of 'Communication' is dealt with in *Health and Safety Guide No. 1 – Human Factors* (Pitman, 1994), a number of features of communication systems must be considered here.

The legal duty of managers to communicate with employees, contractors, visitors and other persons affected by their activities is clearly established in current health and safety legislation, e.g. HSWA, MHSWR. Communication may be officially inspired (formal communication) or unofficial, unplanned and spontaneous (informal communication). It may be one-way, e.g. a direction to operators from a supervisor, or two-way, e.g. where, following the issue of a memorandum, the views of individuals are sought. Two-way communication is more effective in that it gives people the chance to use their intelligence, to contribute knowledge, to participate in the decision-making process, to fulfil their creative needs and to express agreement or disagreement. It helps both the sender and the receiver to measure their standard of achievement and when they both see that they are making progress, their joint commitment to the task will be greater. Communication systems can transmit information upwards, downwards and sideways within an organisation on a one-way or two-way basis.

### Communication on health and safety issues

A manager should make sure he informs his employees of what they need to know. He should not leave it to them to 'read his mind' or 'pick up' the necessary facts. He should ensure they are promptly and accurately informed of matters relevant to their work, including health and safety arrangements and requirements.

Training sessions play an important part in communication. The following matters needs careful consideration if training activities are to be successful and get the appropriate messages over to staff.

1. A list of topics to be covered should be developed, followed by the formulation of a specific training programme.
2. Training sessions should be frequent but should not last longer than 30 minutes.
3. Extensive use should be made of visual aids – videos, films, slides, flip charts, etc.

4. Topics should, as far as possible, be of direct relevance to the group, e.g. the introduction of a safe system of work.
5. Participation should be encouraged with a view to identifying possible misunderstandings or concerns that people may have. This is particularly important when introducing a new safety procedure, such as a permit to work system, or operating procedure.
6. Consideration must be given to eliminating any boredom, loss of interest or adverse response from participants. Sessions should operate on as friendly and informal a basis as possible and in a relatively relaxed atmosphere. Many people respond adversely to the formal classroom situations commonly encountered in many training exercises.

## *Policy and procedure documentation*

Before health and safety at work can be efficiently managed, however, it is necessary to document policies and procedures, clearly identify individual responsibilities and accountabilities and, in effect, establish the 'ground rules' to be followed by all concerned – employers, employees, visitors, contractors, etc. Documentation incorporates a number of well-established aspects.

### The company statement of health and safety policy
This document should form the basis for all health and safety-related activity within an organisation. In large multi-site organisations it may be necessary to frame policy at different levels, e.g. main organisation, at divisional and individual unit levels. Such Statements should be subject to continuing review, particularly in the identification of individual responsibilities and accountabilities.

### Health and safety monitoring systems
Considerable emphasis must be placed on the systems for measuring performance against previously agreed standards. A number of formal systems of health and safety monitoring – safety audits, inspections, health and safety reviews – must be operated, with the appointed competent persons using such systems to assess performance on a continuing basis.

### Objectives and standards
The organisation, and every manager working within it, should have clearly attainable and measurable health and safety-related performance objectives. Such objectives should relate to documented in-house standards laid down in, perhaps, a company Health and Safety Manual, and past incidents which have resulted in poor health and safety performance.

It is common practice in many organisations to identify health and safety-related objectives in the job descriptions or job specifications of all levels of management and staff. A job description generally incorporates the following elements:

(a)   the purpose of the job and its content, i.e. the principal features of the job and how the objectives are achieved

(b)   an organisation chart showing the lines of responsibility and accountability in relation to the job holders

(c)   specific facts and figures about the job, e.g. the number of people the job holders is responsible for, the number of locations covered by him

(d)   problems faced, e.g. securing the commitment of management and staff, coping with various problems, decision-making requirements

(e)   planning and organising aspects

(f)   the direction received

(g)   working contacts, both within and outside the organisation

(h)   knowledge and experience required to undertake the job successfully

(i)   the principal accountabilities of the job holder, e.g. for advising management, planning and developing training programmes, monitoring performance.

## Information, instruction and training

In any health and safety training undertaken, particular attention should be paid to the human factors aspects of safety performance, in particular the potential for human error associated with personal factors such as attitude, motivation, perception and even personality. One of the fundamental objectives of training is to increase awareness of all concerned to hazards and their individual accountabilities and responsibilities for health and safety.

## Hazard elimination

The establishment of formally-documented safe systems of work is an essential feature of compliance with the HSWA. Safe systems of work should identify the hazards, the precautions necessary on the part of staff, influences on behaviour and the training necessary, at induction and other stages of an employee's progress. A Hazard Reporting System is an essential feature of this process, whereby staff can identify and report on hazards that may arise in a number of situations and circumstances.

## Legal compliance

Compliance with the legal requirements must be seen as an essential feature of management's activities. Managers must be fully aware of legal requirements relating to the business operations under their control, their corporate responsibilities and the risk of their being prosecuted specifically under the corporate liability provisions of the HSWA.

## Information retrieval systems

An important feature of health and safety management is the ability to access health and safety information quickly. This is best achieved through an infor-

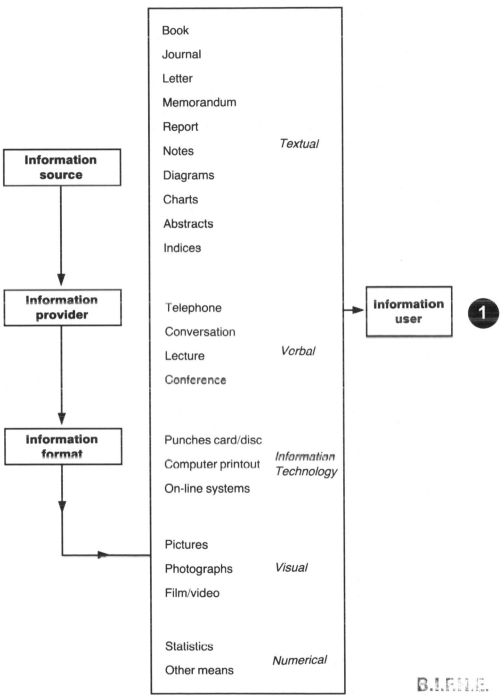

● **FIG 1.2 Information retrieval system**

mation retrieval system. The principal features of such a system are shown in Fig. 1.2: Information retrieval system on the previous page.

## The Management of Health and Safety at Work Regulations 1992

These Regulations came into operation on 1 January 1993 and are covered in more detail in Chapter 2 – *The Legal Duty to Manage Health and Safety at Work.*

They brought in an important new approach to health and safety law, heralding a departure from prescriptive standards of legislation to a more management system-oriented approach by the enforcement agencies. On this basis, enforcement officers, such as factories inspectors and environmental health officers, are now more concerned with the various pro-active strategies being operated by organisations: in particular, the design and operation of safe systems of work, designation and training of 'competent persons', the assessment of risk, health surveillance arrangements, development of emergency procedures and the actual health and safety arrangements in operation. In particular, documentation of procedures and clear evidence of implementation of those procedures are significant.

Greater emphasis will, in future, be placed by the enforcement authorities on the management systems for ensuring good standards of health and safety, with less emphasis on prescriptive standards. The requirement for formally-documented safe systems of work, evidence of risk assessments having been completed and their recommendations implemented, the appointment and training of competent persons and the establishment of emergency procedures is a significant feature of the management process. Managers must appreciate that the room for manoeuvre is drastically reduced under these Regulations in the light of the absolute nature of their requirements. It will not be possible to plead, when charged with an offence, that it was not 'reasonably practicable' in the circumstances under consideration.

# 2

# The legal duty to manage health and safety at work

## COMMON LAW ASPECTS

Common law is the body of accumulated case law of universal, or common, application formed by the judgements of the courts. Each judgement contains the judge's enunciation of the facts, a statement of the law applying to the case and his *ratio decidendi* or legal reasoning for the finding which he has arrived at. These various judgements are recorded in the series of Law Reports and have thus developed into the body of decided case law which we now have and continue to develop. Common law is recorded for the most part in law reports, underpinned by the doctrine of 'precedent', under which a court is bound to follow earlier decisions of courts of its own level and of the decisions of superior courts.

The common law position on health and safety at work is that employers must take reasonable care to protect their employees from risk of foreseeable injury, disease or death at work. The effect of this requirement is that if an employer knows of a health and safety risk to employees, or ought, in the light of the current state of the art, to have known of the existence of a hazard, he will be liable if an employee is injured or killed or suffers illness as a result of the risk, or if the employer failed to take reasonable care.

An employer's duties as common law were identified in general terms by the House of Lords in *Wilsons & Clyde Coal Co. Ltd* v *English* [1938] AC 57. The common law requires that all employers provide and maintain:

(a) a safe place of work with safe means of access and egress

(b) safe appliances and equipment and plant for doing the work

(c) a safe system for doing the work

(d) competent and safety-conscious personnel.

These duties apply even though an employee may be working away on third party premises, or where an employee has been hired out to another employer, but where the control of the task he is performing remains with the permanent employer. The test of whether an employee has been temporarily 'employed' by another employer is one of 'control'.

## The tort of negligence

A tort is a civil wrong. The common law duties of employers listed above are part of the general law of negligence and, as such, are specific aspects of the duty to take reasonable care. 'Negligence' has been defined as:

(a) the existence of a duty of care owed by the defendant to the plaintiff

(b) breach of that duty

(c) damage, loss or injury resulting from or caused by that breach.

(*Lochgelly Iron & Coal Co. Ltd.* v *M'Mullan* [1934] AC 1)

These three facts must be established by an injured employee before he is entitled to bring a claim for damages, although, in the case of breach of statutory duty ((b) above), he merely has to show that the breach of that duty was the material cause of his injury. It is essential that the breach of duty caused the injury or occupational disease (causation). Breach of duty need not be the exclusive cause, but it must be a substantial one, i.e. materially contribute to the injury or ill-health condition.

The legal procedure involving the tort of negligence is for a plaintiff to sue a defendant for damages. If the suit is successful, the plaintiff would be awarded such damages.

There have been a number of important cases involving negligence and its three components, as follows:

### 1. Duty of care

*Donaghue v Stevenson (1932)*
In this case Lord Atkin said: `You must take reasonable care to avoid acts and omissions which you reasonably foresee would be likely to injure your neighbour, i.e. persons who are so closely and directly affected by your act, that you ought reasonably to have them in contemplation as being so affected . . .'

Thus:

(a) there must be a close and direct relationship between the defendant and the plaintiff, e.g. employer and workman, occupier and visitor

(b) the defendant must be able to foresee a real risk of injury to the plaintiff if he, the defendant, does not conduct his operations, or manage his property, with due care.

In this case, the plaintiff bought a bottle of ginger beer which had been manufactured by the defendant. The bottle was of dark glass, so that the contents could not have been seen before they were poured out, and it was sealed, so that it could not have been tampered with until it reached the ultimate consumer. When the plaintiff poured out the ginger beer, the remains of a decomposed snail floated up to the top of her glass and she was taken ill, having already drunk part of the contents of the bottle.

### Bourhill v Young (1943)
In this case, a pregnant Edinburgh fishwife saw a motor cyclist involved in an accident caused by his own negligence. She, some distance away and in perfect safety, suffered shock and subsequently had a miscarriage, the child being stillborn. On this basis, she sued the motor cyclist. It was held by the court that she was outside the foreseeable limit of the duty of care and that

*Res ipsa loquitur* – the facts speak for themselves.
In other words, the circumstances of the accident giving rise to the action are such as impute negligence on the part of the defendant, being an event which, if the defendant had properly ordered his affairs, would not have happened. If this plea by a plaintiff is accepted by the court then a presumption of negligence is raised against the defendant. In other words, effectively it is for the defendant to prove the absence of fault rather than for the plaintiff to prove fault.

### 2. Breach of duty
Whether a breach of duty has occurred is for a judge to decide, taking into consideration the following factors:

(a)   the degree of risk to the plaintiff by the defendant's conduct

### Bolton v Stone (1951)
A cricket ball, hit outside the cricket ground, struck an old lady. It was held that such an event was not reasonably foreseeable, in that it had only happened six times in 30 years. The doctrine of 'reasonable foreseeability' is therefore of significance in such cases.

(b)   the seriousness of harm

### Paris v Stepney Borough Council (1951)
This case established the fact that an employer owes a higher duty of care to a one-eyed workman who, in turn, has a greater risk of blindness than a workman with both eyes. On this basis, it could be argued that other 'vulnerable' groups (e.g. young persons, pregnant women), are owed a higher duty of care than other persons.

(c)   the practicability of the precautions

### Latimer v A.E.C. Limited (1953)
A factory was flooded and employees were called in to clear up the mess. As a result an employee was injured. The Court of Appeal ruled that the cost of precautions to the defendant was too great in relation to the degree of harm

likely to arise. Cost is therefore a factor in cases of negligence.

(d)   current knowledge

*Roe v Ministry of Health (1954)*
Here a patient needed an anaesthetic which was contained in a glass ampoule. Phenol had unknowingly contaminated the anaesthetic through invisible cracks in the glass. The dangers of phenol were well known at the time but not the possibility of the glass ampoules developing invisible cracks which could allow contamination to penetrate. In other words, this risk was not current knowledge to the medical profession at the time.

(e)   current professional practice

To state that a particular way of doing something is 'current professional practice' may be good in evidence, but is not conclusive. Other factors must be considered. Generally, however, if a person holds himself out as having a particular skill, e.g. doctor, surveyor, health and safety practitioner, then he will be expected to use that skill, which implies keeping up to date with what is current professional practice.

### 3. Injury/damage sustained
For the defendant to be liable, it must be shown that his conduct was, as a matter of fact, the cause of the plaintiff's injury or any damage sustained by the plaintiff. However, the 'but for' test is often used, i.e. but for the defendant's conduct, would the injury have occurred?

*Barnett v Chelsea Hospital Management Committee*
In this case a man who had swallowed arsenic was brought into a hospital. The doctor on duty failed to examine him even though he complained of violent vomiting. He later died of arsenic poisoning. However, an action by the relatives, based on negligence, failed on the 'but for' test. The question was asked: 'But for the fact that he was not examined, would his life have been saved?' In this case, even if the condition had been correctly diagnosed, there was very little chance of saving his life.

## Employer's defences

There are two defences available to an employer sued for breach of common law, namely assumption of risk or injury (*volenti non fit injuria*) and contributory negligence.

### (a) *Volenti non fit injuria*
This means 'to one who is willing no harm is done'. It is a complete defence and means that no damages will be payable. It applies to the situation where an employee, being fully aware of the risks he is running in not complying with safety instructions, duties, etc., after being exhorted and supervised,

and having received training and instruction in the dangers involved in not following the laid-down safety procedures and statutory duties, suffers injury, disease and/or death as a result. Here it is open to an employer to argue that the employee agreed to run the risk of the injury involved.

Generally, this defence has not succeeded in actions by injured workers, since there is no presumption that employment, and the dangers sometimes inherent in it, is voluntarily accepted by workers; rather, employment is an economic necessity.

### (b) *Contributory negligence*

The Law Reform (Contributory Negligence) Act 1945 provides that where injury is caused by the fault of two or more persons, liability or fault and, in consequence, damages, must be apportioned in accordance with the extent to which both or more were to blame. On this basis, an employer, when sued by an employee as a result of injury, may be able to claim that the employee contributed to his accident through failure to follow a procedure or safe system of work. Much will depend upon the circumstances of the case, however.

## CRIMINAL LAW

A crime is an offence against the State. Criminal liability refers to the responsibilities under statute and the penalties which can be imposed by the criminal courts, i.e. fines, imprisonment and remedial orders. The criminal courts in question are the magistrates' courts, which handle the bulk of health and safety offences, and the Crown Courts, which deal with the more serious health and safety offences. There are appeal procedures to the High Court and beyond to the Court of Appeal and, assuming that leave is given, to the House of Lords.

The principal statute is the Health and Safety at Work etc. Act 1974. This statute gives the Minister power to make Regulations, the more important of these being the Management of Health and Safety at Work Regulations 1992.

*AIMS AT PEOPLE & ACTIVITIES :*

## HEALTH AND SAFETY AT WORK ETC. ACT 1974 (HSWA)

This Act covers 'all persons at work' whether they be employers, employees or self-employed, except domestic workers in private employment. It is aimed at people and their activities, rather than premises and processes.

The legislation includes both the protection of people at work and the prevention of risks to the health and safety of the general public which may arise from work activities. The objectives of the Act are:

● To secure the health, safety and welfare of all persons at work

- To protect others against the risks arising from workplace activities
- To control the obtaining, keeping and use of explosive or highly flammable substances
- To control emissions into the atmosphere of noxious or offensive substances

## Specific duties under the Act

### General duties of employers to their employees (Section 2)

It is the duty of every employer, so far as is reasonably practicable, to ensure the health, safety and welfare at work of all his employees. More particularly, this includes:

(a) the provision and maintenance of plant and systems of work that are, so far as is reasonably practicable, safe and without risks to health

(b) arrangements for ensuring, so far as is reasonably practicable, safety and absence of risks to health in connection with the use, handling, storage and transport of articles and substances

(c) the provision of such information, instruction training and supervision as is necessary to ensure, so far as is reasonably practicable, the health and safety at work of employees

(d) so far as is reasonably practicable as regards any place of work under the employer's control, the maintenance of it in a condition that is safe and without risks to health and the provision and maintenance of means of access to and egress from it that are safe and without such risks

(e) the provision and maintenance of a working environment for his employees that is, so far as is reasonably practicable, safe, without risks to health, and adequate as regards facilities and arrangements for their welfare at work.

Employers must prepare and, as often as is necessary, revise, a written Statement of Health and Safety Policy, and bring the Statement and any revision of it to the notice of all his employees.

Every employer must consult appointed safety representatives with a view to making and maintaining arrangements which will enable him and his employees to co-operate effectively in promoting and developing measures to ensure the health and safety at work of the employees, and in checking the effectiveness of such measures.

### General duties of employers and self-employed to persons other than their employees (Section 3)

Every employer must conduct his undertaking in such a way as to ensure, so far as is reasonably practicable, that persons not in his employment who may be affected are not thereby exposed to risks to their health or safety. Similar duties are imposed on self-employed persons.

Every employer and self-employed person must give to persons (not being his employees) who may be affected by the way in which he conducts his undertaking the prescribed information about such aspects of it as might affect their health and safety.

## General duties of persons concerned with premises, to persons other than their employees (Section 4)

This section imposes on persons duties in relation to those who are not their employees, but use non-domestic premises made available to them as a place of work.

Every person who has, to any extent, control of premises must ensure, so far as is reasonably practicable, that the premises, all means of access thereto or egress therefrom, and any plant or substances in the premises or provided for use there, is or are safe and without risks to health.

## General duty of persons in control of certain premises in relation to harmful emissions into atmosphere (Section 5)

Any person having control of any premises of a class prescribed for the purposes of Section 1(1)(d) must use the best practicable means of preventing the emission into the atmosphere from the premises of noxious or offensive substances, and of rendering harmless and inoffensive such substances as may be so emitted.

## General duties of manufacturers, etc. as regards articles of substances for use at work (Section 6)

Any person who designs, manufactures, imports or supplies any article for use at work:

(a)  must ensure, so far as is reasonably practicable, that the article is so designed and constructed as to be safe and without risks to health when properly used

(b)  must carry out or arrange for the carrying out of such testing and examination as may be necessary to comply with (a) above

(c)  must provide adequate information about the use for which it is designed and has been tested to ensure that, when put to that use, it will be safe and without risks to health.

Any person who undertakes the design or manufacture of any article for use at work must carry out or arrange for the carrying out of any necessary research with a view to the discovery and, so far as is reasonably practicable, the elimination or minimisation of any risks to health or safety to which the design or article may give rise.

Any person who erects or installs any article for use at work must ensure, so far as is reasonably practicable, that nothing about the way it is erected or installed makes it unsafe or a risk to health when properly used.

Any person who manufactures, imports or supplies any substance for use at work:

(a)   must ensure, so far as is reasonably practicable, that the substance is safe and without risks to health when properly used

(b)   must carry out or arrange for the carrying out of such testing and examination as may be necessary

(c)   must take such steps as are necessary to ensure that adequate information about the results of any relevant tests is available in connection with the use of the substance at work.

### General duties of employees at work (Section 7)

It is the duty of every employee while at work:

(a)   to take reasonable care for the health and safety of himself and of other persons who may be affected by his acts or omissions at work

(b)   as regards any duty or requirement imposed on his employer, to co-operate with him so far as is necessary to enable that duty or requirement to be performed or complied with.

### Duty not to interfere with or misuse things provided pursuant to certain provisions (Section 8)

No person shall intentionally or recklessly interfere with or misuse anything provided in the interests of health, safety or welfare in pursuance of any of the relevant statutory provisions.

### Duty not to charge employees for things done or provided pursuant to certain specific requirements (Section 9)

No employer shall levy or permit to be levied on any employee of his any charge in respect of anything done or provided in pursuance of any specific requirement of the relevant statutory provisions.

## LEVELS OF STATUTORY DUTY

Statutory duties give rise to **criminal liability**. There are three distinct levels of statutory duty.

### *'Absolute' requirements*

Where risk of injury or disease is inevitable if safety requirements are not followed, a statutory duty may well be absolute.

The classic instance of an absolute duty is in Reg. 5(1) of the PUWER which states 'Every employer shall ensure that work equipment is so constructed or adapted as to be suitable for the purpose for which it is to be used or provided'. Absolute duties are defined by the terms 'shall' or must.

## *'Practicable' requirements*

A statutory requirement qualified by the term 'so as is practicable' implies that if in the light of current knowledge and invention it is feasible to comply with this requirement, then irrespective of the cost or sacrifice involved, such a requirement must be complied with (*Schwalb* v *Fass H. & Son* (1946) 175 LT 345). 'Practicable' means more than just physically possible, and implies a higher duty of care than a duty qualified by the term 'so far as is reasonably practicable'.

## *'Reasonably practicable' requirements*

A duty qualified by the term 'so far as is reasonably practicable' implies a lower or lesser level of duty than one which is qualified by 'so far as is practicable'. 'Reasonably practicable' is a narrower term, and implies that a computation must be made in which the quantum or risk is set against the sacrifice involved in the measures necessary for averting that risk. If it can be shown that there is a gross disproportion between these two factors, i.e. the risk being insignificant in relation to the sacrifice, then a defendant discharges the onus upon himself (*Edwards* v *National Coal Board* [1949]) 1 AER 743]. All duties under HSWA are qualified by the term 'so far as is reasonably practicable'.

### The reasonable man

But what is a reasonable person? What is it about their behaviour that makes them reasonable, and how do the courts interpret this term? The mythical 'reasonable man' was identified by one judge in the past as being 'the man who travels to work every day on the top deck of the No. 57 Clapham omnibus'. As such the term is flexible and changes with time according to society and the norms prevalent at the time.

This term can be found in Section 7 of the HSWA, i.e. the duty on an employee to take reasonable care for the safety of himself and others, including members of the public, who may foreseeably be affected by his acts or omissions at work.

## THE STATEMENT OF HEALTH AND SAFETY POLICY

Under Section 2(3) of HSWA every employer has a duty to 'prepare and as often as appropriate revise a written statement of his general policy with respect to the health and safety at work of his employees and the organisation and arrangements for the time being in force for carrying out that policy, and to bring the statement and any revision of it to the notice of his employees'.

The Statement of Health and Safety Policy is, fundamentally, the key document for detailing the management systems and procedures to ensure sound levels of health and safety performance. It should be revised at regular intervals, particularly before changes in the structure of the organisation, the introduction of new articles and substances, and changes in legal requirements affecting the organisation. Fundamentally, the Statement must be seen as a living document which reflects the current organisational arrangements, the hazards and precautions necessary, the individual responsibilities of people and the systems for monitoring performance.

## Objectives of the policy statement

- It should affirm long range purpose
- It should commit management at all levels and reinforce this purpose in the decision-making process
- It should indicate the scope left for decision-making by junior managers

## Scope of the policy

A well-written statement of health and safety policy should cover the following aspects:
- Management intent
- The organisation and arrangements for implementing the policy
- Individual accountabilities of directors, line managers, employees and other groups, e.g. contractors
- Details of the organisation with respect to both line and staff functions
- The role and function of health and safety specialists, e.g. health and safety advisers, nursing advisers, occupational hygienists, occupational health nurses, unit safety officers, occupational physicians, company doctors and trade union-appointed safety representatives.

## Principal features of a statement of health and safety policy

A Statement of Health and Safety Policy should incorporate the following features:-

1. A general statement of intent which states the basic objectives, supplemented by details of the organisation and arrangements (rules and procedures).
2. Definition of both the duties and extent of responsibility at specified line management levels for health and safety, with identification made at the highest level of the individual with overall responsibility for health and safety.
3. Definition of the function of the safety adviser/officer and his relationship to senior and line management made clear.
4. The system for monitoring safety performance and publishing of information about that performance.
5. An identification and analysis of hazards together with the precautions necessary on the part of staff, visitors, contractors, etc.

6. An information system which will be sufficient to produce an identification of needs and which can be used as an indicator of effectiveness of the policy.
7. A policy on the provision of information, instruction and training for all levels of the organisation.
8. A commitment to consultation on health and safety and to a positive form of worker involvement.
9. The Statement should bear the signature of the person with ultimate responsibility and accountability for health and safety at work, e.g. chief executive, managing director.

## MANAGEMENT OF HEALTH AND SAFETY AT WORK REGULATIONS 1992 (MHSWR)

These Regulations came into operation on 1 January 1993 and are accompanied by an ACOP issued by the Health and Safety Commission. The duties, because of their wide-ranging general nature, overlap with many existing Regulations, e.g. Control of Substances Hazardous to Health Regulations 1988 (COSHH). Where duties overlap, compliance with the duty in the more specific regulation will normally be sufficient to comply with the corresponding duty in the MHSWR. However, where the duties in these regulations go beyond those in the more specific regulations, additional measures will be needed to comply fully with the MHSWR.

Because of their importance in the areas of health and safety management, the principal requirements of the Regulations are dealt with below.

### Regulation 1 – Interpretation

*The assessment* in the case of an employer, the assessment made by him in accordance with Reg. 3(1) and changed by him where necessary in accordance with Reg. 3(3); and in the case of a self-employed person, the assessment made by him in accordance with Reg. 3(2) and changed by him in accordance with Reg. 3(3).

*Employment business* a business (whether or not carried on with a view to profit and whether or not carried on in conjunction with any other business) which supplies persons (other than seafarers) who are employed in it to work for and under the control of other persons in any capacity.

*Fixed-term contract of employment* a contract of employment for a specific term which is fixed in advance or which can be ascertained in advance by reference to some relevant circumstance.

*The preventive and protective measures* the measures which have been identified by the employer or by the self-employed person in consequence of the assessment as the measures he needs to take to comply with the requirements and prohibitions imposed upon him by or under the relevant statutory provisions.

## Regulation 2 – Disapplication of these Regulations

These Regulations shall not apply to or in relation to the master or crew of a seagoing ship or to the employer of such persons in respect of the normal shipboard activities of a ship's crew under the direction of the master.

## Regulation 3 – Risk assessment

1. Every employer shall make a suitable and sufficient assessment of:
   (a) the risks to the health and safety of his employees to which they are exposed whilst at work
   (b) the risks to the health and safety of persons not in his employment arising out of or in connection with the conduct by him of his undertaking, for the purpose of identifying the measures he needs to take to comply with the requirements and prohibitions imposed upon him by or under the relevant statutory provisions

2. Similar provisions as in 1 above apply in the case of self-employed persons.

3. Any assessment shall be reviewed by the employer or self-employed person who made it if
   (a) there is reason to suspect it is no longer valid
   (b) there has been a significant change in the matters to which it relates

   Where, as a result of any such review, changes to an assessment are required, the employer or self-employed persons shall make them.

4. Where the employer has five or more employees, he shall record:
   (a) the significant findings of the assessment
   (b) any group of his employees identified by it as being especially at risk

Risk assessment is the principal feature of these Regulations. See Chapter 3 – *Risk and Risk Assessment*.

## Regulation 4 – Health and safety arrangements

1. Every employer shall make and give effect to such arrangements as are appropriate, having regard to the nature of his activities and the size of his undertaking, for the effective planning, organisation, control, monitoring and review of the preventive and protective measures.

2. Where the employer has five or more employees, he shall record these arrangements.

There is a need here to consider the systems necessary to ensure effective management of health and safety requirements. Such systems should be integrated with other management systems, e.g. those for financial, personnel, production, engineering, purchasing and other areas of management activity. The principal elements of management practice, i.e. planning, organising, controlling, monitoring and reviewing, should be taken into account.

## *Regulation 5 – Health surveillance*

Every employer shall ensure that his employees are provided with such health surveillance as is appropriate, having regard to the risks to their health and safety which are identified by the assessment.

Health surveillance may already be required in order to comply with existing legislation, such as the COSHH. However, it may also be necessary where the risk assessment under these Regulations indicates that:

(a) there is an identifiable disease or adverse health condition related to the work concerned

(b) valid techniques are available to detect indications of the disease or condition, e.g. certain forms of biological monitoring

(c) there is a reasonable likelihood that the disease or condition may occur under the particular conditions of work

(d) surveillance is likely to further the protection of the health of the employees to be covered

The principal objective of any health surveillance is to detect adverse health effects at an early stage, thereby enabling further harm to be prevented.

## *Regulation 6 – Health and safety assistance*

1. Every employer shall, subject to paras 6 and 7, appoint one or more competent persons to assist him in undertaking the measures he needs to take to comply with the requirements and prohibitions imposed upon him by or under the relevant statutory provisions.
2. Where an employer appoints persons in accordance with para 1, he shall make arrangements for ensuring adequate co operation between them.
3. The employer shall ensure that the number of persons appointed under para 1, the time available for them to fulfil their functions, and the means at their disposal are adequate, having regard to the size of his undertaking, the risks to which his employees are exposed and the distribution of those risks throughout the undertaking.
4. The employer shall ensure that:
    (a) any person appointed by him in accordance with para 1 who is not in his employment
        (i) is informed of the factors known by him to affect, or suspected by him of affecting, the health and safety of any other person who may be affected by the conduct of his undertaking
        (ii) has access to the information referred to in reg. 8.
    (b) any person appointed by him in accordance with para 1 is given such information about any person working in his undertaking who is
        (i) employed by him under a fixed-term contract of employment,

or

(ii) employed in an employment business, as necessary to enable that person properly to carry out the function specified in that paragraph

5. A person shall be regarded as competent for the purposes of para 1 where he has sufficient training and experience or knowledge and other qualities properly to undertake the measures referred to in that paragraph

6. Para 1 shall not apply to a self-employed employer who is not in partnership with any other person, where he has sufficient training and experience or knowledge and other qualities properly to undertake the measures referred to in that paragraph.

7. Para 1 shall not apply to individuals who are employers and who are together carrying on business in partnership where at least one of the individuals concerned has sufficient training and experience or knowledge and other qualities:

(a) properly to undertake the measures he need to take to comply with the requirements and prohibitions imposed upon him by or under the relevant statutory provisions

(b) properly to assist his fellow partners in undertaking the measures they need to take to comply with the requirements and prohibitions imposed upon them by or under the relevant statutory provisions

The concept of 'competent persons' is not new to health and safety legislation, the appointment of such persons being required under the Construction Regulations, Noise at Work Regulations, Pressure Systems and Transportable Gas Containers Regulations, etc. The degree of competence required of the competent person for the purposes of these Regulations will depend on the risks identified in the risk assessment. While there is no specific emphasis on the qualifications of such persons, broadly a competent person should have such skill, knowledge and experience as to enable him to identify hazards and understand the implications of those hazards. The depth of training, accountabilities and responsibilities, authority and level of reportability within the management system of the competent person is, therefore, important. The competent person should not be placed so far down the management system that his recommendations carry no weight.

See also *Competent persons* on page 36.

## Regulation 7 – Procedures for serious and imminent danger and for danger areas

1. Every employer shall:

(a) establish and, where necessary, give effect to appropriate procedures to be followed in the event of serious and imminent danger to persons at work in his undertaking

(b) nominate a sufficient number of competent persons to implement those procedures in so far as they relate to the evacuation from premises of persons at work in his undertaking

(c) ensure that none of his employees has access to any area occupied by him to which it is necessary to restrict access on ground of health and safety unless the employee concerned has received adequate health and safety instruction

2. Without prejudice to the generality of para 1(a), the procedures referred to in that sub-paragraph shall:

(a) so far as is practicable, require any persons at work who are exposed to serious and imminent danger to be informed of the nature of the hazard and of the steps to be taken to protect them from it

(b) enable the persons concerned (if necessary by taking appropriate steps in the absence of guidance or instruction and in the light of their knowledge and the technical means at their disposal) to stop work and immediately proceed to a place of safety in the event of their being exposed to serious, imminent or unavoidable danger

(c) save in exceptional cases for reasons duly substantiated (which cases and reasons shall be specified in those procedures), require the persons concerned to be prevented from resuming work in any situation where there is still a serious and imminent danger

3. A person shall be regarded as competent for the purposes of para 1(b) where he has sufficient training and experience or knowledge and other qualities to enable him properly to implement the evacuation procedures referred to in that sub-paragraph.

The requirement here is that employers must establish procedures to be followed by workers if situations present serious and imminent danger, and under what circumstances they should stop work and move to a place of safety. Fundamentally, organisations' emergency plans/procedures must cover foreseeable high risk situations, such as fire, explosion, etc., and ensure training of staff in these procedures.

## Regulation 8 – Information for employees

Every employer shall provide his employees with comprehensible and relevant information on:

(a) the risks to their health and safety identified by the assessment

(b) the preventive and protective measures

(c) the procedures referred to in reg. 7(1)(a)

(d) the identity of those persons nominated by him in accordance with reg. 7(1)(b)

(e) the risks notified to him in accordance with reg. 9(1)(c)

The significant feature of this Regulation is that information provided to workers must be 'comprehensible', i.e. written in such a way as to be easily understood by the people to which it is addressed. On this basis, the mode of presentation of such information should take account of their level of training, knowledge and experience. It may also need to consider people with language difficulties or with disabilities which may impede their receipt of information. Information can be provided in whatever form is most suitable in the circumstances, so long as it is comprehensible, e.g. staff handbook, posters.

## Regulation 9 – Co-operation and co-ordination

1. Where two or more employers share a workplace (whether on a temporary or permanent basis) each such employer shall:

   (a) co-operate with the other employers concerned so far as is necessary to enable them to comply with the requirements and prohibitions imposed upon them by or under the relevant statutory provisions

   (b) (taking into account the nature of his activities) take all reasonable steps to co-ordinate the measures he takes to comply with the requirements and prohibitions imposed upon him by or under the relevant statutory provisions with the measures the other employers concerned are taking to comply with the requirements and prohibitions imposed upon them by or under the relevant statutory provisions

   (c) take all reasonable steps to inform the other employers concerned of the risks to their employees' health and safety arising out of or in connection with the conduct by him of his undertaking

2. Para 1 shall apply to employers sharing a workplace with self-employed persons and to self-employed persons sharing a workplace with other self-employed persons as it applies to employers sharing a workplace with other employers; and the references in that paragraph to employers and the reference in the said paragraph to their employees shall be construed accordingly.

This Regulation makes provision for employers jointly occupying a work site, e.g. a construction site, trading estate or office block, to co-operate and co-ordinate their health and safety activities, particularly where there are risks common to everyone, irrespective of their work activity. Fields of co-operation and co-ordination include health and safety training, safety monitoring procedures, emergency arrangements, welfare amenity provisions, etc. In some cases, the appointment of a health and safety co-ordinator should be considered.

## Regulation 10 – Persons working in host employers' or self-employed persons' undertakings

1. Every employer and every self-employed person shall ensure that the employer of any employees from an outside undertaking who are working in his undertaking is provided with comprehensible information on:

   (a) the risks to those employees' health and safety arising out of or in connection with the conduct by that first-mentioned employer or by that self-employed person of his undertaking

   (b) the measures taken by the first-mentioned employer by that self-employed person in compliance with the requirements and prohibitions imposed upon him by or under the relevant statutory provisions in so far as the said requirements and prohibitions relate to those employees

2. Para 1 shall apply to a self-employed person who is working in the undertaking of an employer or a self-employed person as it applies to employees from an outside undertaking who are working therein; and the reference in that paragraph to the employer of any employees from an outside undertaking who are working in the undertaking of an employer or a self-employed person and the references in the said paragraph to employees from an outside undertaking who are working in the undertaking of an employer or a self-employed person shall be construed accordingly.

3. Every employer shall ensure that every person working in his undertaking who is not his employee and every self-employed person (not being an employer) shall ensure that any person working in his undertaking is provided with appropriate instructions and comprehensible information regarding any risks to that person's health and safety which arise out of the conduct by that employer or self-employed person of his undertaking.

4. Every employer shall:

   (a) ensure that the employer of any employees from an outside undertaking who are working in his undertaking is provided with sufficient information to enable that second-mentioned employer to identify any person nominated by that first-mentioned employer in accordance with Reg. 7(1)(b) to implement evacuation procedures as far as those employees are concerned

   (b) take all reasonable steps to ensure that any employees from an outside undertaking who are working in his undertaking receive sufficient information to enable them to identify any person nominated by him in accordance with Reg. 7(1)(b) to implement evacuation procedures as far as they are concerned

5. Para 4 shall apply to a self-employed person who is working in an employer's undertaking as it applies to employees from an outside undertaking who are working therein; and the reference in that paragraph to the

employer of any employees from an outside undertaking who are working in an employer's undertaking and the references in the said paragraph
to employees from an outside undertaking who are working in an
employer's undertaking shall be construed accordingly.

This Regulation applies to a wide range of work activities where employees
fundamentally undertake work in other people's premises, e.g. contract clean-
ers, maintenance staff, etc. Such persons must be provided with adequate information and instructions regarding relevant risks to their health and safety.

## Regulation 11 – Capabilities and training

1. Every employer shall, in entrusting tasks to his employees, take into
   account their capabilities as regards health and safety.
2. Every employer shall ensure that his employees are provided with adequate health and safety training:
   (a) on their being recruited into the employer's undertaking
   (b) on their being exposed to new or increased risks because of
       (i)   their being transferred or given a change of responsibilities
             within the employer's undertaking
       (ii)  the introduction of new work equipment or a change
             respecting work equipment already in use within the
             employer's undertaking
       (iii) the introduction of new technology into the employer's
             undertaking
       (iv)  the introduction of a new system of work or a change
             respecting a system of work already in use within the
             employer's undertaking
3. The training referred to in para 2 shall:
   (a) be repeated periodically where appropriate
   (b) be adapted to take account of new or changed risks to the health and
       safety of the employees concerned
   (c) take place during working hours

This Regulation introduces, for the first time in health and safety legislation,
'a human factors' approach to ensuring appropriate levels of health and safety provision. When allocating work to employees, employers should ensure
that the demands of the job do not exceed the employees' ability to carry out
the work without risk to themselves and others. There is a need, therefore,
for employers to consider both the physical and mental abilities of employees before allocating tasks, together with their level of knowledge, training
and experience. Training, in particular, should reflect this aspect of human
capability.

## Regulation 12 – Employees' duties

1. Every employee shall use any machinery, equipment, dangerous sub-stances, transport equipment, means of production or safety device pro-vided to him by his employer in accordance both with any training in the use of the equipment concerned which has been received by him and the instructions respecting that use which have been provided to him by the said employer in compliance with the requirements and prohibitions imposed upon that employer by or under the relevant statutory provi-sions.

2. Every employee shall inform his employer or any other employee of that employer with specific responsibility for health and safety of his fellow employees:

   (a) of any work situation which a person with the first-mentioned employee's training and instruction would reasonably consider rep-resented a serious and immediate danger to health and safety

   (b) of any matter which a person with the first-mentioned employee's training and instruction would reasonably consider represented a shortcoming in the employer's protection arrangements for health and safety, in so far as that situation or matter either affects the health and safety of that first-mentioned employee or arises out of or in con-nection with his own activities at work, and has not previously been reported to his employer or to any other employee of that employer in accordance with this paragraph.

This Regulation reinforces and expands the duties of employers towards employees under Section 7 of the HSWA. In the light of these requirements employers should install some form of hazard reporting system, whereby employees can report hazards to their employer or appointed competent person. Such a system should ensure a prompt response where hazards are identified, with the competent person signing off the hazard report when the hazard has been eliminated or controlled.

## Regulation 13 – Temporary workers

1. Every employer shall provide any person whom he has employed under a fixed-term contract of employment with comprehensible information on:

   (a) any special occupational qualifications or skills required to be held by that employee if he is to carry out his work safely

   (b) any health surveillance required to be provided to that employee by or under any of the relevant statutory provisions, and shall provide the said information before the employee concerned commences his duties

2. Every employer and every self-employed person shall provide any person employed in an employment business who is to carry out work in his

undertaking with similar comprehensible information as indicated in 1(a) and (b) above.

3. Every employer and every self-employed person shall ensure that every person carrying on an employment business whose employees are to carry out work in his undertaking is provided with comprehensible information on:

(a)  any special occupational qualifications or skills required to be held by those employees if they are to carry out their work safely

(b)  the specific features of the jobs to be filled by those employees (in so far as those features are likely to affect their health and safety)

and the person carrying on the employment business concerned shall ensure that the information so provided is given to the said employees.

This Regulation supplements previous Regulations requiring the provision of information, with additional requirements on temporary workers, i.e. those employed on fixed duration contracts and those employed in employment businesses, but working under the control of a user company. The use of temporary workers will also have to be notified to the competent person.

## Regulation 14 – Exemption certificates

These Regulations may, by certificate of the Secretary of State for Defence, exclude members of the forces generally.

## Regulation 15 – Exclusion of civil liability

Breach of a duty imposed by these Regulations shall not confer a right of action in any civil proceedings. (Similar provisions apply in the case of HSWA.)

## The schedule

The following regulation shall be inserted after regulation 4 of the Safety Representatives and Safety Committee Regulations 1977:

*Employer's duty to consult and provide facilities and assistance*

4A. 1. Without prejudice to the generality of Section 2(6) of the Health and Safety at Work Act 1974, every employer shall consult safety representatives in good time with regard to:

(a)  the introduction of any measure at the workplace which may substantially affect the health and safety of the employees the safety representatives concerned represent

(b)  his arrangements for appointing or, as the case may be, nominating persons in accordance with regs. 6(1) and 7(1)(b) of the MHSWR 1992

(c)     any health and safety information he is required to provide to the employees the safety representatives concerned represent by or under the relevant statutory provisions

(d)     the planning and organisation of any health and safety training he is required to provide to the employees' safety representatives concerned represent by or under the relevant statutory provisions

(e)     the health and safety consequences for the employees the safety representatives concerned represent of the introduction (including planning thereof) of new technologies into the workplace

2. Without prejudice to regulations 5 and 6 of these Regulations, every employer shall provide such facilities and assistance as safety representatives may reasonably require for the purpose of carrying out their functions under Section 2 of the 1974 Act and under these Regulations'.

### Approved Code of Practice
These Regulations are supported by a comprehensive Approved Code of Practice issued by the Health and Safety Commission.

### Footnote to the MHSWR
It must be appreciated by managers and others that the vast majority of the duties specified in these Regulations are of an absolute nature, designated by the term 'shall', compared with the HSWA, where the duties are qualified by the term 'so far as is reasonably practicable', a lower level of duty.

## CORPORATE LIABILITY

Under HSWA directors, managers, company secretaries and similar officers of the body corporate have both general and specific duties. Breaches of these duties can result in individuals being prosecuted.

### Offences committed by companies (Section 37)(1)

Where a breach of one of the relevant statutory provisions on the part of a body corporate is proved to have been committed with the consent or connivance of, or to have been attributable to any neglect on the part of, any director, manager, secretary or other similar officer of the body corporate or a person who was purporting to act in any such capacity, he as well as the body corporate shall be guilty of that offence and shall be liable to be proceeded against and punished accordingly.

Breach of this section has the following consequences:
- Where an offence is committed through neglect by a board of directors, the company itself can be prosecuted as well as the directors individually who may have been to blame.

- Where an individual functional director is guilty of an offence, he can be prosecuted as well as the company.
- A company can be prosecuted even though the act or omission was committed by a junior official or executive or even a visitor to the company.

Generally, most prosecutions under Section 37(1) would be limited to that body of persons, i.e. the board of directors and individual functional directors, as well as senior managers.

### Offences committed by other corporate persons (Section 36)

Section 36 makes provision for dealing with offences committed by corporate officials, e.g. personnel managers, health and safety specialists, training officers, etc. Thus:

Where the commission by any person of an offence under any of the relevant statutory provisions is due to the ACT or DEFAULT of some other person, that other person shall be guilty of the offence, and a person may be charged with and convicted of the offence by virtue of this subsection whether or not proceedings are taken against the first mentioned person.

## COMPETENT PERSONS

One way of ensuring the operation of a safe system of work is by the designation and employment of specifically trained operators who appreciate the risks involved. This may be for undertaking certain inspections, issuing permits to work and supervising their operation, or carrying out work where there is a high degree of foreseeable risk.

The expression 'competent person' occurs frequently in construction safety law. For example, under the Construction (General Provisions) Regulations 1966 certain inspections, examinations, operations and supervisory duties must be undertaken by competent persons. However, it should be noted that 'competent persons' are not generally defined in law except in the Electricity at Work Regulations 1989 and the Pressure Systems and Transportable Gas Containers Regulations 1989 (see below). Therefore the onus is on the employer to decide whether persons are competent to undertake these duties. An employer might do this by reference to the person's training, qualifications and experience. Broadly, a competent person should have practical and theoretical knowledge as well as sufficient experience of the particular machinery, plant or procedure involved to enable him to identify defects or weaknesses during plant and machinery examinations, and to assess their importance in relation to the strength and function of that plant and machinery (*Brazier* v *Skipton Rock Company Ltd.* [1962] 1 AER 955).

Competent persons are involved in a wide range of activities under specific legislation, e.g. the various Construction Regulations, Noise at Work Regulations 1989 and Pressure Systems and Transportable Gas Containers

Regulations 1989, as shown in the following extracts. Their role and function is frequently identified in ACOPs .

## Power Presses Regulations 1965

Under Reg. 4 of these Regulations no person must:

(a)  set, re-set, adjust or try out tools on a power press

(b)  install or adjust any safety device

(c)  carry out an inspection and test of a safety device, unless:
    (i)  he is 18 years or over
    (ii)  he has been trained
    (iii) he is competent to perform the duties
    (iv) he has been appointed by the factory occupier to undertake the above duties.

## Construction Regulations

Competent persons must be involved in the following:

(a)  supervision of demolition work

(b)  supervision of the handling and use of explosives

(c)  inspection of scaffold materials prior to erection

(d)  supervision of the erection of, substantial alteration or additions to, and the dismantling of scaffolds

(e)  inspection of scaffolds every 7 days and following adverse weather conditions which could affect the strength and stability of a scaffold, or cause displacement of any part

(f)  inspection of excavations on a daily basis

(g)  supervision of the erection of cranes

(h)  testing of cranes after erection, re-erection and any removal or adjustment involving a change of anchorage or ballasting

(j)  examination of appliances for anchorage or ballasting prior to the erection of a crane.

## Offshore Installations (Operational Safety, Health and Welfare) regulations 1976

Regulation 6. requires that every **lifting appliance** and **every piece of** lifting gear be thoroughly examined and tested by a competent person:

(a)  before use for the first time

(b)  if already used, when substantially altered or repaired

(c)   in the case of examination, before putting into use after
    (i)    installation
    (ii)   re-installation
    (iii)  substantial alteration or repair and at 6 monthly intervals

(d)   in the case of testing before being put into use after
    (i)    installation
    (ii)   re-installation
    (iii)  substantial alteration or repair

Regulation 23 requires that there must be at any time when an installation is manned, at least:

(a)   one person fully trained as a radio-telephone operator

(b)   a competent person responsible for the control of helicopter operations whose function is to ensure, before a helicopter lands or takes off, that:
    (i)    the helicopter landing area is clear of obstructions
    (ii)   any cranes nearby have ceased to operate
    (iii)  no persons, other than those necessary, are in the helicopter landing area
    (iv)   fire-fighting equipment, manned by adequately trained persons, is available
    (v)    any vessel standing by to render assistance is informed that helicopter operations are to take place
    (vi)   safety nets are properly secured

## Noise at Work Regulations 1989

Regulation 4 requires that every employer shall, when any of his employees is likely to be exposed to the first action level or above, or to the peak action level or above, ensure that a competent person makes a noise assessment which is adequate for the purposes:

(a)   of identifying which of his employees are so exposed

(b)   of providing him with such information with regard to the noise to which those employees may be exposed as will facilitate compliance with his duties under Reg. 7 (reduction of exposure to noise), Reg. 8 (provision of ear protectors), Reg. 9 (marking of, and restriction of entry to, ear protection zones) and Reg. 11 (provision of information, instruction and training).

## Pressure Systems and Transportable Gas Containers Regulations 1989

Under these Regulations owners and users of pressure systems must have a written scheme of examination drawn up by a competent person for the examination of the system at specified intervals to confirm that the system continued to comply with the Regulations and that there is no risk of danger due to the uncontrolled release of stored energy from any part of the system.

The term 'competent person' is dealt with in the ACOP, and the degree of competence, in terms of qualifications, experience, specialist services available and structure of the competent person's organisation, is specified according to the type of work undertaken, i.e. on minor systems, intermediate systems or major systems as defined. Generally, however, the term 'competent person' is used in connection with three distinct functions:

(a) advising the user on the scope of the written scheme

(b) drawing up or certifying schemes of examination

(c) undertaking examinations under the scheme

It is the responsibility of users to select a competent person capable of undertaking these duties in a proper manner.

The ACOP does specify that 'any individual, when carrying out competent person duties, should be sufficiently independent from the interests of all other functions to ensure adequate segregation of accountabilities'. This is particularly appropriate where the appointed person is appointed from within an organisation which uses pressure systems.

## Electricity at Work Regulations 1989

Regulation 16 requires that no person carry out a work activity where technical knowledge or experience is necessary to prevent danger or injury, unless he has such knowledge or experience or is under the appropriate degree of supervision.

## Management of Health and Safety at Work Regulations 1992

These Regulations bring in important new provisions relating to the appointment and mode of operation of 'competent persons', both generally to ensure the employer is complying with legal requirements (Reg. 6), and specifically in connection with 'procedures for serious and imminent dangers and for danger areas' (Reg. 7). The ACOP accompanying the MHSWR makes the following points:

1. Employers must have access to competent help in applying the provisions of health and safety law, including these Regulations and in particular to devising and applying protective measures, unless they are competent to undertake the measures without assistance. Appointment of competent persons for this purpose should be included among the arrangements recorded under Regulation 4(2).

2. Employers may appoint one or more of their own employees to do all that is necessary, or may enlist help or support from outside the organisation,

or they may do both. Employers who are sole traders, or are members of partnerships, may appoint themselves (or other partner) to carry out health and safety measures. Large employers may well appoint a whole department with specific health and safety responsibilities including specialists in such matters as occupational hygiene or safety engineering. In any case where external support is brought in, its activities must be co-ordinated by those appointed by the employer to manage the health and safety measures.

3. External services employed will usually be appointed in an advisory capacity only. They will often be specialists or general consultants on health and safety matters.

4. The appointment of such health and safety assistants, departments or advisers does not absolve the employer from responsibilities for health and safety under the Health and Safety at Work Act and other relevant statutory provisions. It can do no more than give added assurance that these responsibilities will be discharged adequately.

5. Employers are solely responsible for ensuring that those they appoint to assist them with health and safety measures are competent to carry out whatever tasks they are assigned and given adequate information and support. In making these decisions employers should take into account the need for:

   (a)  a knowledge and understanding of the work involved, the principles of risk assessment and prevention, and current health and safety applications

   (b)  the capacity to apply this to the task required by the employer, which might include identifying the health and safety problems, assessing the need for action, designing and developing strategy and plans, implementing these strategies and plans, evaluating their effectiveness and promoting and communicating health and safety and welfare advances and practices.

6. Competence in the sense it is used in these Regulations does not necessarily depend on the possession of particular skills or qualifications. Simple situations may require only the following:

   (a)  an understanding of relevant current best practice

   (b)  awareness of the limitations of one's own experience and knowledge

   (c)  the willingness and ability to supplement existing experience and knowledge

7. The provision of effective health and safety measures in more complex or highly technical situations will call for specific applied knowledge and skills which can be offered by appropriately qualified specialists. In the case of specific knowledge and skills in occupational health and safety, membership of a professional body or similar organisation at an appropriate level and in an appropriate part of health and safety, can help to guide employers. Competence-based qualifications accredited by the

National Council for Vocational Qualifications and SCOTVEC (the Scottish Vocational Education Council), which are being developed for most occupations, may also provide a guide.

## HEALTH AND SAFETY ADVISERS

THE HSE publication 'Successful Health and Safety Management' makes the following points with regard to the role and functions of health and safety advisers, (be they internally-appointed persons or external consultants):

Organisations that successfully manage health and safety give health and safety advisers the status and ensure they have the competence to advise management and workers with authority and independence. Subjects on which they advise include:

(a) health and safety policy formulation and documentation

(b) structuring and operating all parts of the organisation (including the supporting system) in order to promote a positive health and safety culture and to secure the effective implementation of policy (see Chapter 10 – *Developing a safety culture*)

(c) planning for health and safety, including the setting of realistic short and long-term objectives, deciding priorities and establishing adequate performance standards

(d) day-to-day implementation and monitoring of policy and plans, including accident and incident investigation, reporting and analysis

(e) reviewing performance and auditing the whole safety management system

To fulfil these functions, they have to:

(a) maintain adequate information systems in relevant law (civil and criminal) and on guidance and developments in general and safety management practice

(b) be able to interpret the law and understand how it applies to the organisation

(c) establish and keep up to date organisational and risk control standards relating to both 'hardware' (such as the place of work and the plant, substances and equipment in use) and 'software' (such as the procedures, system and people); this task is likely to involve contributions from specialists, for example, architects, engineers, doctors and occupational hygienists

(d) establish and maintain procedures for the reporting, investigating and recording and analysis of accidents and incidents

(e) establish and maintain adequate and appropriate monitoring and auditing systems

(f)   present themselves and their advice in an independent and effective manner, safeguarding the confidentiality of personal information such as medical records.

## Relationships of health and safety advisers

The position of health and safety advisers in the organisation is such that they support the provision of authoritative and independent advice. The post holder has a direct reporting line to directors on matters of policy and authority to stop work which is being carried out in contravention of agreed standards and which puts people at risk of injury. Health and safety advisers have responsibility for professional standards and systems and on a large site or in a group of companies may also have line management responsibilities for junior health and safety professionals.

They are also involved in liaison with a wide range of outside bodies and individuals, including local authority environmental health officers and licensing officials, architects and consultants, etc., the fire service, contractors, insurance companies, clients and customers, the HSE, the public, equipment suppliers, HM Coronor or Procurator Fiscal, the media, the police, general practitioners and hospital staff.

## SOURCES OF INFORMATION

Health and safety at work is a multi-disciplinary subject, which requires an understanding of many disciplines – law, engineering, occupational health and hygiene, ergonomics, human factors, etc. To manage health and safety effectively, good sources of information are essential. These can be classified as follows:

## Formal (primary) sources

### 1. E.C. Directives
These are the Community instrument of legislation. Directives are legally binding on the governments of all member states who must introduce national legislation, or use administrative procedures where applicable, to implement its requirements.

### 2. U.K. Acts of Parliament (statutes)
Acts of Parliament can be innovatory, i.e. introducing new legislation, or consolidating, i.e. reinforcing, with modifications, existing law. Statutes empower the Minister or Secretary of State to make Regulations (delegated or subordinate legislation). Typical examples are the HSWA and the Factories Act 1961, the latter consolidating and reinforcing the Factories Act 1937.

### 3. Regulations (statutory instruments)

Regulations are more detailed than the parent' Act, which lays down the framework and objectives of the system. Specific details are incorporated in Regulations made under the Act, e.g. the COSHH Regulations 1988 which were passed pursuant to the HSWA. Regulations are made by the appropriate Minister or Secretary of State whose powers to do so are identified in the parent Act.

### 4. Statutory Rules and Orders

These are the earlier equivalent of Statutory Instruments. They ceased to be published in 1948, but there are many applicable to occupational health and safety which are still in force, e.g. Chains, Ropes and Lifting Tackle (Register) Order 1938.

### 5. Approved Codes of Practice

The HSC is empowered to approve and issue Codes of Practice for the purpose of providing guidance on health and safety duties and other matters laid down in statute or Regulations. A Code of Practice can be drawn up by the Commission or the Health and Safety Executive. In every case, however, the relevant government department, or other body, must be consulted beforehand and approval of the Secretary of State must be obtained. Any Code of Practice approved in this way is an Approved Code of Practice (ACOP).

An ACOP enjoys a special status under the HSWA. Although failure to comply with any provision of the code is not in itself an offence, that failure may be taken by a court, in criminal proceedings, as proof that a person has contravened the legal requirement to which the provision relates. In such a case it will be open to that person to satisfy a court that he or she has complied with the requirement in some other way.

Examples of ACOPs are those issued with the COSHH Regulations entitled 'Control of Substances Hazardous to Health', 'Control of Carcinogenic Substances' and 'Control of Substances Hazardous to Health in Fumigation Operations'.

### 6.  Case law (common law)

Case law is an important source of information. It is also known as common law because, traditionally, judges have formulated rules and principles of law as the cases occur for decision before the courts.

What is important in a case is the *ratio decidendi* (the reason for the decision). This is binding on courts of equal rank who may be deciding the same point of law. *Ratio decidendi* is the application of such an established principle to the facts of a given case, for instance, negligence consists of omitting to do what a 'reasonable man' would do in order to avoid causing injury to others.

Case law is found in law reports, for example, the All England Law Reports, the Industrial Cases Reports, the current Law Year Book and in professional journals, e.g. Law Society Gazette, Solicitors Journal. In addition, many newspapers carry daily law reports e.g. *The Times, Financial Times, Daily Telegraph* and *The Independent.* The supreme law court is the European Court of Justice, whose decisions are carried in *The Times.*

## Non-legal sources

A wide range of non-legal sources of information is available, some of which, however, may be quoted in legal situations.

### 1. HSE Series of Guidance Notes
Guidance Notes issued by the HSE have no legal status. They are issued on a purely advisory basis to provide guidance on good health and safety practices, specific hazards, etc. There are five series of Guidance Notes – General, Chemical Safety, Plant and Machinery, Medical and Environmental Hygiene.

### 2. British standards
These are produced by the British Standards Institute. They provide sound guidance on numerous issues and are frequently referred to by enforcement officers as the correct way of complying with a legal duty. For instance, British Standard 5304 'Safeguarding of machinery' is commonly quoted in conjunction with the duties of employers under Sec. 14 of the Factories Act 1961 to provide and maintain safe machinery.

### 3. Manufacturers' information and instructions
Under Section 6 of the HSWA (as amended by the Consumer Protection Act 1987) manufacturers, designers, importers and installers of 'articles and substances used at work' have a duty to provide information relating to the safe use, storage, etc. of their products. Such information may include operating instructions for machinery and plant, and hazard data sheets in respect of dangerous substances. Information provided should be sufficiently comprehensive and understandable to enable a judgement to be made on their safe use at work.

### 4. Safety organisations
Safety organisations, such as the Royal Society for the Prevention of Accidents (RoSPA) and the British Safety Council, provide information in the form of magazines, booklets and videos on a wide range of health and safety-related topics.

### 5. Professional institutions
Many professional institutions, such as the Institute of Occupational Safety and Health, the Institution of Environmental Health Officers, the British Occupational Hygiene Society, etc., provide information, both verbally and in written form.

### 6. Published information
This takes the form of textbooks, magazines, law reports, updating services, microfiche systems, films and videos on general and specific topics.

## *Internal information sources*

There are many sources of information available within organisations. These include:

### 1. Existing written information
This may take the form of Statements of Health and Safety Policy, company Health and Safety Codes of Practice, specific company policies, e.g. on the use and storage of dangerous substances, current agreements with trade unions, company rules and regulations, methods, operating instructions, etc. Such documentation could be quoted in a court of law as an indication of the organisation's intention to regulate activities in order to ensure legal compliance. Evidence of the use of such information in staff training is essential here.

### 2. Work study techniques
Included here are the results of activity sampling, surveys, method study, work measurement and process flows.

### 3. Job descriptions
A job description should incorporate health and safety responsibilities and accountabilities. It should take account of the physical and mental requirements and limitations of certain jobs and any specific risks associated with the job. Representations from operators, supervisors and trade union safety representatives should be taken into account. Compliance with health and safety requirements is an implied condition of every employment contract, breach of which may result in dismissal or disciplinary action by the employer.

### 4. Accident statistics
Statistical information on past accidents and sickness may identify unsatisfactory trends in operating procedures which can be eliminated at the design stage of safe systems of work. The use of accident statistics and rates, e.g. accident incidence rate, as a sole measure of safety performance is not recommended, however, due to the variable levels of accident reporting in work situations. Under-reporting of accidents, common in many organisations, can result in inaccurate comparisons being made between one location and another.

### 5. Task analysis
Information produced by the analysis of tasks, such as the mental and physical requirements of a task, manual operations involved, skills required,

influences on behaviour, hazards specific to the task, and learning methods necessary to impart task knowledge must be taken into account. Job safety analysis, a development of task analysis, will provide the above information prior to the development of safe systems of work.

### 6. Direct observation

This is the actual observation of work being carried out. It identifies interrelationships between operators, hazards, dangerous practices and situations and potential risk situations. It is an important source of information in ascertaining whether, for instance, formally designed safe systems of work are being operated or safety practices, imparted as part of former training activities, are being followed.

### 7. Personal experience

People have their own unique experience of specific tasks and the hazards which those tasks present. The experiences of accident victims, frequently recorded in accident reports, are an important source of information. Feedback from accidents is crucial in order to prevent repetition of them.

### CONCLUSION

Health and safety law embraces both the common law and the criminal law. The HSWA imposes a wide range of duties on employers and managers, and this Act has been substantially reinforced by the absolute requirements of the MHSWR and other recent legislation. This means, fundamentally, that the room for manoeuvre on the part of employers, when charged with offences under health and safety law, has been greatly reduced.

There is now a legal duty on the part of organisations to manage their health and safety activities and to show clear-cut documentary evidence of their management systems. Evidence of documented health and safety procedures, together with clear evidence of these procedures being put into practice, will increasingly be required by the courts in order to show compliance with the absolute requirements of the MHSWR and other Regulations.

# Risk and risk assessment

## THE LEGAL DUTY TO ASSESS RISKS

Risk assessment is the basis of the Regulations which came into operation on 1 January 1993. However, this duty is not new. Health risk assessments are required under the COSHH Regulations 1988 and hearing risk assessments under the Noise at Work Regulations 1989. The duties under the 1992 'six pack' legislation are outlined below.

### *Management of Health and Safety at Work Regulations 1992 (MHSWR)*

These Regulations require an employer to assess:

(a) the risks to the health and safety of his employees to which they are exposed whilst at work
(b) the risks to the health and safety of persons not in his employment arising out of or in connection with the conduct by him of his undertaking

for the purpose of identifying the measures he needs to take to comply with his duties under the relevant statutory provisions (Reg. 3(1)). He must also revise this assessment when it is no longer valid because of new or changed risks (Reg 3(3)).

### *Provision and Use of Work Equipment Regulations 1992 (PUWER)*

Regulation 5 of these Regulations covers the suitability of work equipment thus:

1. Every employer shall ensure that work equipment is so constructed or adapted as to be suitable for the purpose for which it is used or provided.
2. In selecting work equipment, every employer shall have regard to the

working conditions and to the risks to the health and safety of persons which exist in the premises or undertaking in which that work equipment is to be used, and any additional risk posed by the use of that work equipment.

3. Every employer shall ensure that work equipment is used only for operations for which, and under conditions for which, it is suitable.

4. In this regulation 'suitable' means suitable in any respect which it is reasonably foreseeable will affect the health or safety of any persons.

### Approved Code of Practice

62. The risk assessment carried out under Regulation 3(1) of the Management of Health and Safety at Work Regulations 1992 will help employers to select work equipment and assess its suitability for particular tasks.

## Personal Protective Equipment at Work Regulations 1992 (PPEWR)

Under these Regulations an employer must provide suitable personal protective equipment (Reg. 4). Regulation 6 requires an assessment to be made before choosing any particular personal protective equipment thus:

1. Before choosing any personal protective equipment which by virtue of regulation 4 he is required to ensure is provided, an employer or self-employed person shall ensure that an assessment is made to determine whether the personal protective equipment he intends will be provided is suitable.

2. The assessment required under paragraph 1 shall include:

   (a) an assessment of any risk or risks to health or safety which have not been avoided by other means

   (b) the definition of the characteristics which personal protective equipment must have in order to be effective against the risks referred to in sub-paragraph (a) of this paragraph, taking into account any risks which the equipment itself may create

   (c) comparison with the characteristics of the personal protective equipment available with the characteristics referred to in sub-paragraph (b) of this paragraph

3. Every employer or self-employed person who is required by paragraph 1 to ensure that any assessment is made shall ensure that any such assessment is reviewed if:

   (a) there is reason to suspect that it is no longer valid

   (b) there has been a significant change in the matters to which it relates and where as a result of any such review changes in the assessment are required, the relevant employer or self-employed person shall ensure that they are made.

**Guidance.**

37. The purpose of the assessment provision in Regulation 6 is to ensure that the employer who needs to provide PPE chooses PPE which is correct for the particular risks involved and for the circumstances of its use. It follows from, but does not duplicate, the risk assessment requirements of the Management of Health and Safety at Work Regulations 1992.

## Manual Handling Operations Regulations 1992 (MHOR)

Regulation 4 of these Regulations is concerned with the duties of employers thus:

1. Each employer shall;
   (a) so far as is reasonably practicable, avoid the need for his employees to undertake any manual handling operations at work which involve a risk of their being injured
   (b) where it is not reasonably practicable to avoid the need for his employees to undertake any manual handling operations at work which involve a risk of their being injured:
      (i) make a suitable and sufficient assessment of all such manual handling operations to be undertaken by them, having regard to the factors which are specified in column 1 of Schedule 1 to these Regulations and considering the questions which are specified opposite thereto in column 2 of that Schedule;
      (ii) take appropriate steps to reduce the risk of injury to those employees arising out of their undertaking any such manual handling operations to the lowest level reasonably practicable;
      (iii) take appropriate steps to provide any of those employees who are undertaking any such manual handling operations with general indications and, where it is reasonably practicable to do so, precise information on:
         (aa) the weight of each load, and
         (bb) the heaviest side of any load whose centre of gravity is not positioned centrally.

2. Any assessment such as is referred to in paragraph (1)(b)(i) of this regulation shall be reviewed by the employer who made it if –
   (a) there is reason to suspect it is no longer valid
   (b) there has been a significant change in the manual handling operations to which it relates

   and where as a result of any such review changes to an assessment are required, the relevant employer shall make them.

## Health and Safety (Display Screen Equipment) Regulations 1992 (HSDSER)

These Regulations require an employer to undertake workstation analysis in order to reduce risks to display screen equipment users. Regulation 2 is framed thus:

1. Every employer shall perform a suitable and sufficient analysis of those workstations which:

   (a) (regardless of who has provided them) are used for the purposes of his undertaking by users

   (b) have been provided by him and are used for the purposes of his undertaking by operators

   for the purpose of assessing the health and safety risks to which those persons are exposed in consequence of that use.

2. Any assessment made by an employer in pursuance of paragraph (1) shall be reviewed by him if:

   (a) there is reason to suspect that it is no longer valid

   (b) there has been a significant change in the matters to which it relates

   and where as a result of any such review changes to an assessment are required, the employer concerned shall make them.

3. The employer shall reduce the risks in consequence of an assessment to the lowest extent reasonably practicable.

4. The reference in paragraph (3) to 'an assessment' is a reference to an assessment made by the employer concerned in pursuance of paragraph (1) and changed by him where necessary in pursuance of paragraph (2).

Clearly, no two people assess risk in the same way, let alone attempt to quantify it. To comply with the Regulations, organisations need some form of assessment procedure that will be acceptable, in a court in particular. Furthermore, the system for measuring and evaluating risk should be acceptable to all parties concerned – employers, employees, enforcement agencies and, perhaps, insurers.

Accurate risk assessment is perhaps best undertaken using a team approach – manager, safety adviser, safety representative, specific employees – the team having been trained in a standardised approach to such assessment and acting within prescribed parameters. Inaccurate risk assessment will result from individual managers undertaking their own uncoordinated assessments. Examples of risk assessment documentation are shown in Figs. 3.1–3.7 at the end of this chapter.

## RISK, HAZARD AND DANGER – THE DISTINCTION

Various definitions of the term **risk** are available, e.g.

*'A chance of loss or injury.'*

*'A chance of bad consequences.'*

*'To expose to a hazard'.*

*'To venture.'*

*'To expose to a mischance.'*

*'To expose to chance of injury or loss.'*

*'The probability of harm, damage or injury.'*

*'The probability of a hazard leading to personal injury and the severity of that injury.'*

A **hazard**, on the other hand, is defined as:

*'A situation of risk or danger.'*

*'A situation that may give rise to personal injury.'*

*'The result of a departure from the normal situation, which has the potential to cause injury, damage or loss'.*

**Danger** is defined as *'liability or exposure to harm; something that causes peril'.*
Research shows that no two people perceive risk in the same way, such perception being based on human factors such as attitude to danger, sensory perception, motivation, personality, past experience and the level of arousal, together with individual skill, knowledge and experience .

## RISK ASSESSMENT PRINCIPLES

Risk assessment is not a precise science. The assessment of risk fundamentally considers a number of factors, namely the likelihood or probability that an accident or incident could be caused, the severity of the outcome, in terms of injury, damage or loss and the number of people affected, and the frequency of exposure to risk. A simple method of assessing it is through the use of risk ratings taking into account the factors of probability, severity and frequency on a scale from 1 to 10 in each case. (Some systems use only probability and severity factors.)

**Risk Rating = Probability (P) x Severity (s) x Frequency (F)**

which gives a rating between 1 and 1000. Standard probability, severity and frequency indices are used (see Tables 1, 2 and 3). The urgency or priority of action with regard to a particular risk associated with a task or activity can be quantified as shown in Table 4 – Priority of action scale.

**Table 1   Probability scale**

| Probability Index | Descriptive phrase |
|---|---|
| 10 | Inevitable |
| 9 | Almost certain |
| 8 | Very likely |
| 7 | Probable |
| 6 | More than even chance |
| 5 | Even chance |
| 4 | Less than even chance |
| 3 | Improbable |
| 2 | Very improbable |
| 1 | Almost impossible |

**Table 2   Severity scale**

| Severity Index | Descriptive phrase |
|---|---|
| 10 | Death |
| 9 | Permanent total incapacity |
| 8 | Permanent severe incapacity |
| 7 | Permanent slight incapacity |
| 6 | Absent from work for more than 3 weeks with subsequent recurring incapacity |
| 5 | Absent from work for more than 3 weeks but with subsquent complete recovery |
| 4 | Absent from work for more than 3 days but less than 3 weeks with subsequent complete recovery |
| 3 | Absent from work for less than 3 days with complete recovery |
| 2 | Minor injury with no lost time and complete recovery |
| 1 | No human injury expected |

**Table 3   Frequency scale**

| Frequency Index | Descriptive phrase |
| --- | --- |
| 10 | Hazard permanently present |
| 9 | Hazard arises every 30 seconds |
| 8 | Hazard arises every minute |
| 7 | Hazard arises every 30 minutes |
| 6 | Hazard arises every hour |
| 5 | Hazard arises every shift |
| 4 | Hazard arises once a week |
| 3 | Hazard arises once a month |
| 2 | Hazard arises every year |
| 1 | Hazard arises every 5 years |

**Table 4   Priority of action scale**

| | |
| --- | --- |
| 800 – 1000 | Immediate action |
| 000 – 800 | Action within next 7 days |
| 400 – 600 | Action within next month |
| 200 – 400 | Action within next year |
| Below 200 | No immediate action necessary, but keep under review |

## Alternative form of risk assessment

A simple risk assessment formula involving frequency, severity and maximum possible loss can also be used thus:

**Risk Rating = Frequency x (Severity + MPL + Probability)**

where Severity = Number of people at risk

Frequency = Frequency of the occurrence

Maximum possible loss is based on a scale from 1 to 50

The criteria for assessing risk in this way are shown in Tables 5 and 6 respectively, together with Priority of Action in Table 7.

**Table 5   Typical scale of maximum possible loss**

| | |
|---|---|
| 50 | Fatal |
| 45 | Loss of two limbs/eyes |
| 40 | Loss of hearing |
| 30 | Loss of one limb/eye |
| 15 | Broken arm |
| 1 | Scratch |

**Table 6  Alternative probability scale**

| | |
|---|---|
| 50 | Imminent |
| 35 | Hourly |
| 25 | Daily |
| 15 | Once a week |
| 10 | Once a month |
| 5 | Once a year |
| 1 | Unlikely |

**Table 7  Alternative priority of action**

| Risk Rating | Urgency of action |
|---|---|
| Over 100 | Immediate |
| 50 – 100 | Today |
| 25 – 50 | Within 1 week |
| 10 – 25 | Within 1 month |
| 1– 10 | Within 3 months |

## WHAT IS A RISK ASSESSMENT?

A risk assessment may be defined as:

> **an identification of the hazards present in an undertaking and an estimate of the extent of the risks involved, taking into account whatever precautions are already being taken.**

It is essentially a four-stage process:

(a)  identification of all the hazards

(b)  measurement of the risks

(c)  evaluation of the risks

(d)  implementation of measures to eliminate or control the risks.

There are different approaches which can be adopted in the workplace, e.g.

(a)  examination of each activity which could cause injury

(b)  examination of hazards and risks in groups e.g. machinery, substances, transport; and/or

(c)  examination of specific departments, sections, offices, construction sites.

In order to be *suitable* and *sufficient* and to comply with legal requirements, a risk assessment must:

(a)  identify all the hazards associated with the operation, and evaluate the risks arising from those hazards, taking into account current legal requirements

(b)  record the significant findings if more than five persons are employed, even if they were in different locations

(c)  identify any employee or group of employees who are especially at risk

(d)  identify others who may be specially at risk, e.g. visitors, contractors, members of the public

(e)  evaluate existing controls, stating whether or not they are satisfactory and, if not, what action should be taken

(f)  evaluate the need for information, instruction, training and supervision

(g)  judge and record the probability or likelihood of an accident occurring as a result of uncontrolled risk, including the 'worst case' likely outcome

(h)  record any circumstances arising from the assessment where serious and imminent danger could arise

(i)  provide an action plan giving information on implementation of additional controls, in order of priority, and with a realistic timescale.

The assessment must be recorded where more than five persons are employed (electronic methods of recording are acceptable). The assessment must incorporate details of items (a) to (j) above.

## Generic assessments

There are assessments produced once only for a given activity or type of workplace. In cases where an organisation has several locations or situations where the same activity is undertaken then a generic risk assessment could be carried out for a specific activity to cover all locations. Similarly, where operators work away from the main location and undertake a specific task, e.g. installation of telephones or servicing of equipment, a generic assessment should be produced. For generic assessments to be effective:

(a) 'worst case' situations must be considered

(b) provision should be made on the assessment to monitor implementation of the assessment controls which are/are not relevant at a particular location, and what action needs to be taken to implement the relevant required actions from the assessment.

In certain cases, there may be risks which are specific to one situation only, and these risks may need to be incorporated in a separate part of the generic risk assessment.

## Maintaining the risk assessment

The risk assessment must be maintained. This means that any significant change to a workplace, process or activity, or the introduction of any new process, activity or operation, should be subject to risk assessment. If new hazards come to light, then these should also be subject to risk assessment. The risk assessment should in any case be periodically reviewed and updated. This is best achieved by a suitable combination of safety inspection and monitoring techniques, which require corrective and/or additional action where the need is identified. Typical monitoring systems include:

(a) preventive maintenance inspections

(b) safety representative/committee inspections

(c) statutory and maintenance scheme inspections, tests and examinations

(d) safety tours and inspections

(e) occupational health surveys

(f) air monitoring

(g) safety audits.

Useful information on checking performance against control standards can also be obtained reactively from the following activities:

(a) accident and ill-health investigation

(b) investigation of damage to plant, equipment and vehicles

(c) investigation of 'near miss' situations.

The frequency of review depends upon the level of risk in the operation, and should normally be at least every ten years. Further, if a serious accident occurs in the organisation, or elsewhere but is possible in the organisation, and where a check on the risk assessment shows no assessment or a gap in assessment procedures, then a review is necessary.

## *Risk/hazard control*

Once the risk or hazard has been identified and assessed, employers must either prevent the risk arising or, alternatively, control it. Much will depend upon the magnitude of the risk in terms of the controls applied. In certain cases, the level of competence of operators may need to be assessed before they undertake certain work, e.g. work on electrical systems.

A typical hierarchy of control, from high risk to low risk, is indicated below.

1. *Elimination* of the risk completely, e.g. prohibiting a certain practice or the use of a certain hazardous substance.
2. *Substitution* by something less hazardous or risky.
3. *Enclosure* of the risk in such a way that access is denied.
4. *Guarding or the installation of safety devices* to prevent access to danger points or zones on work equipment and machinery.
5. *Safe systems of work* that reduce the risk to an acceptable level.
6. *Written procedures*, e.g. job safety instructions, that are known and understood by those affected.
7. *Adequate supervision* , particularly in the case of young or inexperienced persons.
8. *Training* of staff to appreciate the risks and hazards.
9. *Information*, e.g. safety signs, warning notices.
10. *Personal protective equipment*, e.g. eye, hand, head and other forms of body protection.

In many cases, a combination of the above control methods may be necessary. It should be appreciated that the amount of management control necessary will increase proportionately for the controls lower down this list. In other words, item 1 indicates no control is needed, whereas item 10 requires a high degree of management control.

It may be necessary to consider the following commonly-occurring hazards when undertaking risk assessments.

| | |
|---|---|
| Fall of a person from a height | Contact with hot/cold surfaces |
| Fall of an object/material from a height | Compressed air |
| Fall of a person on the same level | Mechanical lifting operations |
| Manual handling | Noise and vibration |
| Use of work equipment | Biological agents |
| Operation of vehicles | Radiation |
| Fire | Adverse weather |
| Electricity | Hazardous substances |
| Drowning | Storage of goods |
| Excavation work | Housekeeping/cleaning |
| Stored energy | Temperature, lighting and ventilation |
| Explosions | |

## Risk assessment documentation

Whilst risk assessment is a standard feature of much recent legislation, there is no standard format or layout in terms of their actual documentation. Risk assessments may relate to a particular workplace or work activity (see Fig. 3.1 – Workplace risk assessment and Fig. 3.2 – Work activity risk assessment). Or they may be more specific and detailed, relating to, for instance, work equipment, personal protective equipment and display screen equipment (see Figs. 3.5, 3.6 and 3.7 respectively).

The principal objective of any risk assessment document is to identify the key risks, to quantify those risks, and to identify information requirements, responsibility for action and the record-keeping requirements. The Risk Assessment Summary (see Fig. 3.3) should specify, fundamentally, the extent of the risk – high, medium or low – and the precautions to be taken by all persons exposed to that risk. A Risk Assessment Action Plan should be produced and implemented as a result of the risk assessment exercise.

---

## WORKPLACE RISK ASSESSMENT

No _____    Date _____    Valid unit _____

1.  Department/Work Area

2.  Work activities in the workplace

3.  Key Risks

4.  Risk assessment (based on identified hazards)

| Risk Rating |
|---|
|  |

    4.1   Fire protection
    4.2   Emergency procedure
    4.3   Vehicle movements
    4.4   Electrical installations
    4.5   Pressure systems
    4.6   Welfare amenity provisions
    4.7   Environmental factors
    4.8   Articles and substances
    4.9   Falls and falling objects

5.  Information and authorisation
      Emergency procedures
   – Training standards
   – Safe systems of work
   – Safety notices
   – Reference documents

6.  Record-keeping requirements

Risk assessment summary

Date _____    Assessor _____

---

● **FIG 3.1  Workplace risk assessment**

---

# WORK ACTIVITY RISK ASSESSMENT

No _____ Date _____ Valid unit _____

1.  Department/Work Area

2.  Work activitiy

    Responsible person

3.  Key risks

4.  Risk assessment (based on identified hazards)

    | Risk Rating |
    | --- |
    | |

    4.1  Fire protection
    4.2  Machinery
    4.3  Display screen equipment
    4.4  Manual handling
    4.5  Hazardous substances
    4.6  Noise and vibration
    4.7  Human factors
    4.8  Environmental factors
    4.9  Mobile handling equipment
    4.10 Electrical installations

5.  Information and authorisation
    – Emergency procedures
    – Training standards
    – Safe systems of work
    – Safety notices
    – Reference documents

6.  Record-keeping requirements

Risk assessment summary

Date _____ Assessor _____

---

● **FIG 3.2 Work activity risk assessment**

**RISK ASSESSMENT SUMMARY**

**Risk Assement No.**

Work activity/Workplace _____

Date of assessment _____ Assessor _____

Principal risks

Specific risks

**3**

**Remedial Action**
1. Immediate

2. Short term (28 days)

3. Medium term (6 months)

4. Long term (over 12 months)

**Information, Instruction and Training Requirements**

**Supervision Requirements**

Date of next review _____

● **FIG 3.3 Risk assessment summary**

**RISK ASSESSMENT ACTION PLAN**

| ACTIVITY/SITUATION HAZARDS | ACTION REQUIRED | TARGET DATE | ACTION BY | COMPLETED |
|---|---|---|---|---|
| | | | | |

This Action Plan prepared by

Date _____        Next Assessment before _____

● **FIG 3.4  Risk assessment action plan**

# WORK EQUIPMENT RISK ASSESSMENT

Item of work equipment _____

Plant Register No _____

Operations for which the work equipment is used _____

_____

Frequency of use – Regular use/Irregular use

Date of Assessment _____ Assessor _____

| | Yes/No | Level of Risk | | |
|---|---|---|---|---|
| | | Low | Med | High |
| **Suitability (Reg. 5)** | | | | |
| 1. It is constructed or adapted to be suitable for the purpose for which it is used. | | | | |
| 2. It was selected taking into account working conditions, health risks and any additional risks posed by its use. | | | | |
| 3. It is used only for operations for which, and under conditions for which, it is suitable. | | | | |
| **Maintenance (Reg. 6)** | | | | |
| 4. It is maintained in an efficient state, in efficient working order and in good repair. | | | | |
| 5. The maintenance log, where used, is kept up to date. | | | | |
| **Specific risks (Reg. 7)** | | | | |
| 6. Where it is likely to involve a specific risk– | | | | |
| (a) its use is restricted to designated trained users; and | | | | |
| (b) any repairs, modifications or servicing of same are undertaken by specifically designated persons. | | | | |
| **Information and Instructions (Reg. 8)** | | | | |
| 7. Adequate health and safety information and, where appropriate, written instructions, are available to users, managers and supervisors. | | | | |

● **FIG 3.5 Work equipment risk assessment**

| | Yes/No | Level of Risk | | |
|---|---|---|---|---|
| | | Low | Med | High |

8. The information and, where appropriate, the written instructions for use, include–
   (a) the conditions in which, and the methods by which it may be used;
   (b) foreseeable abnormal situations and the action to be taken if such a situation were to occur; and
   (c) any conclusions to be drawn from experience in using same.

9. Information and instructions are readily comprehensible to users of the equipment.

**Training (Reg. 9)**
10. All persons who use it, managers and supervisors, have received adequate training in the methods which may be adopted when using it, any risks which its use may entail and precautions to be taken.

**Dangerous parts of machinery (Reg. 11)**
11. In the case of machinery, effective measures are taken –
    (a) to prevent access to any part of machinery or to any rotating stock bar
    (b) to stop the movement of any dangerous part of machinery or rotating stock-bar before any person enters a danger zone.

12. The above measures consist of –
    (a) provisions of fixed guards; or
    (b) provision of other guards or protection devices; or
    (c) provision of jigs, holders, push sticks or similar protection appliances used in conjunction with the machinery; or

● **FIG 3.5 Continued**

| | Yes/No | Level of Risk | | |
|---|---|---|---|---|
| | | Low | Med | High |

(d)  provision of information, instruction, training and supervision.

13.  All guards and protection devices –

(a)  are suitable for their purpose;
(b)  are of good construction, sound material and adequate strength;
(c)  are maintained in an efficient state, in efficient working order and in good repair;
(d)  do not give rise to any increased risk to health or safety;
(e)  are not easily bypassed or disabled;
(f)  are situated at sufficient distance from any danger zone;
(g)  do not unduly restrict the view of the operating cycle of the machinery, where such a view is necessary;
(h)  are so constructed or adapted that they allow operations necessary to fit or replace parts and for maintenance work, restricting access so that it is allowed only to the area where the work is not to be carried out and, if possible, without having to dismantle the guard or protection device.

**Specified hazards (Reg. 12)**
14.  Measures listed below are taken to ensure that the exposure of the user to any of the undermentioned risks to his health or safety is either prevented or adequately controlled, namely –

(a)  measures other than the provision of personal protective equipment or of information, instruction, training and supervision;
(b)  include, where appropriate, measures to minimise the effects of the hazard occuring, with particular reference to –

**3**

● **FIG 3.5  Continued**

| | Yes/No | Level of Risk | | |
|---|---|---|---|---|
| | | Low | Med | High |

(i)   any article or substance falling or being ejected from the work equipment;

(ii)  rupture or disintegration of parts of the work equipment;

(iii) the work equipment catching fire or overheating;

(iv) the unintended or premature discharge of any article or of any gas, dust, liquid, vapour or other substance which, in each case, is produced, used or stored in the work equipment

(v)  the unintended or premature explosion of the work equipment, or any article or substance produced, used or stored in it.

**High or very low temperature (Reg. 13)**

15. The work equipment, parts of same and any article or substance produced, used or stored in same which, in each case, is at a high or very low temperature has protection where appropriate so as to prevent injury to any person by burn, scald or sear.

**Controls for starting or making a significant change in operating conditions (Reg. 14)**

16. Where appropriate, it is provided with one or more controls for the purpose of –

(a)  starting same, incuding re-starting after a stoppage; or

(b)  controlling any change in the speed, pressure or other operating conditions where such conditions after the change result in risk to health or safety which is greater than or of a different nature from such risks before the change.

17. Where the above control is required, it is not possible to perform any operation above except by a deliberate action on such control.

● **FIG 3.5 Continued**

| | Yes/No | Level of Risk | | |
|---|---|---|---|---|
| | | Low | Med | High |

**Stop controls (Reg. 15)**

18. Where appropriate, it is provided with one or more readily accessible controls the operation of which will bring same to a safe condition in a a safe manner.

19. Any above control brings it to a complete stop where necessary for reasons of health or safety.

20. If necessary for reasons of health or safety, any above control switches of all sources of energy after stopping the functioning of the equipment.

21. The above control operates in priority to any control which stars or changes the operating conditions of the equipment.

**Emergency stop controls (Reg. 16)**

22. Where appropriate, it is provided with one or more readily accessible emergency stop controls unless it is not necessary by reason of the nature of the hazards and the time taken for the work equipment to come to a complete stop as a result of the action of any stop control.

23. The control operates in priority to a stop control.

**Controls (Reg. 17)**

24. The controls are clearly visible and identifiable, including by appropriate marking where necessary.

25. No control is in a position where any person operating the control is exposed to a risk.

26. Where appropriate –
    (a) the operator of any control is able to ensure from the position of that control that no person is in a place where he could be exposed to a risk as a result of the operation of that control; or

**3**

● **FIG 3.5 Continued**

| | Yes/No | Level of Risk | | |
|---|---|---|---|---|
| | | Low | Med | High |

(b)  systems of work are effective to ensure that, when the equipment is about to start, no person is in a place where he would be exposed to risk; or

(c)  an audible, visible or other suitable warning is given whenever the equipment is about to start.

27.  Appropriate measures are taken to ensure that any person who is in a place where he would be exposed to risk as a result of the starting or stopping of work equipment has sufficient time and suitable means to avoid that risk.

**Control systems (Reg. 18)**

28.  All control systems are safe in that –
 (a)  their operation does not create any increased risk;
 (b)  they ensure that any fault in or damage to any part of the control system or the loss of supply of any source of energy cannot result in additional or increased risk; and
 (c)  they do not impede the operation stop control or emergency stop control.

**Isolation from sources of energy (Reg. 19)**

29.  Where appropriate, the equipment is provided with suitable clearly identifiable and readily accessible means to isolate it from all sources of energy.

30.  Appropriate measures are taken to ensure that reconnection of any energy source does not expose the user to risk.

**Stability (Reg. 20)**

31.  Where necessary, the equipment, or any part of same, is stabilised by clamping.

● **FIG 3.5 Continued**

| | Yes/No | Level of Risk | | |
|---|---|---|---|---|
| | | Low | Med | High |

**Lighting (Reg. 21)**

32. Suitable and sufficient lighting, which takes account of the operations to be carried out is provided at the places where the equipment is used.

**Maintenance and operations (Reg. 22)**

33. Appropriate measures are taken to ensure the equipment is so constructed or adapted that maintenance operations which involve a risk can be carried out while the equipment is shut down or, in other cases –

    (a) maintenance operations can be carried out without exposing the maintenance operator to risk; or

    (b) appropriate measures can be taken for the protection of any maintenance operator exposed to a risk.

**Markings (Reg. 23)**

34. The equipment is marked in a clearly visible manner with any marking appropriate for reasons of health and safety.

**Warnings (Reg. 24)**

35. The equipment incorporates any warnings or warning devices which are appropriate for reasons of health and safety.

36. Warnings given by warning devices are unambiguous, easily perceived and easily understood.

● **FIG 3.5 Continued**

**RISK ASSESSMENT SUMMARY**

**Principal Risks**

**Specific Risks**

**Remedial Action**

1.  Immediate

2.  Short term (28 days)

3.  Medium term (6 months)

4.  Long term (over 12 months)

**Maintenance Requirements**

**Information, Instruction, Training and Supervision Requirements**

**Date of next review**

● **FIG 3.5  Continued**

## PERSONAL PROTECTIVE EQUIPMENT
## RISK ASSESSMENT

Item of personal protective equipment (PPE) _____

Risk exposure situation(s) for which the PPE is provided

_____

Frequency of use – Regular use/Irregular use

Date of Assessment _____ Assessor _____

| | Yes/No | Level of Risk | | |
|---|---|---|---|---|
| | | Low | Med | High |

**Provision and suitability (Reg. 4)**

1. The PPE is provided on the basis that the risks have not been controlled by other means which are equally or more effective.

2. It is suitable in that:–
   (a) it is appropriate to the risk or risks involved and the conditions at the place where exposure to the risk may occur:
   (b) it takes account of ergonomic requirements and the state of health of the persons who may wear it;
   (c) it is capable of fitting the wearer correctly, if necessary, after adjustments within the range for which it is designed; and
   (d) so far as is practicable, it is effective to prevent or adequately control the risks involved without increasing overall risk.

**Compatibility (Reg. 5)**

3. Where this item of PPE is used with other items of PPE, such equipment is compatible and continues to be effective against the risks in question.

● **FIG 3.6 Personal protective equipment risk assessment**

| | Yes/No | Level of Risk | | |
|---|---|---|---|---|
| | | Low | Med | High |

**Assessment (Reg. 6)**

4. The following matters have been considered
   to determine whether the PPE  intended
   to be provided is suitable, including:–

   (a) risks which have not been avoided by
       other means;
   (b) the definition of the characteristics
       which this PPE must have in order to be
       effective against the above risks,
       taking into account any risks which the
       PPE itself may create; and
   (c) comparison of the characteristics of the
       PPE with the characteristics in (b) above.

5. The following specific risks have been
   assessed, namely, risk of –

   (a) head injury;
   (b) eye and/or face injury;
   (c) inhalation of airborne contaminants;
   (d) noise-induced hearing loss;
   (e) skin contact:
   (f) bodily injury;
   (g) hand and arm injury;
   (h) leg and foot injury; and
   (i) vibration-induced injury.

**Maintenance and replacement of PPE (Reg. 7)**

6. Procedures are in operation to ensure the
   PPE is maintained (including replacement)
   or cleaning as appropriate) in an
   efficient state, in efficient working
   order and in good repair

● **FIG 3.6 Continued**

| | Yes/No | Level of Risk | | |
|---|---|---|---|---|
| | | Low | Med | High |

**Accommodation for PPE (Reg. 8)**

7.  Appropriate accommodation is provided
    for the PPE when not in use.

**Information, instruction and training (Reg. 9)**

8.  Employees using this PPE have been provided
    with such information, instructions and
    training as is adequate and appropriate to
    enable the employee to know –

    (a)  the risk(s) which the PPE will avoid or limit;
    (b)  the purpose and the manner in which
         the PPE is to be used; and
    (c)  any action to be taken by the employee
         to ensure the PPE remains in efficient
         state, in efficient working order and
         in good repair.

**Use of PPE**

9.  Procedures operate to ensure PPE is properly
    used by employees.

10. Employees use the PPE  provided in accordance
    with the training  received by them and the
    instructions respecting that use.

11. Employees take all reasonable steps to ensure
    the PPE is returned to the accommodation
    after use.

**Reporting loss or defect**

12. Employees understand the need to report
    forthwith any loss or obvious defect in the
    PPE to their employer.

● **FIG 3.6 Continued**

**RISK ASSESSMENT SUMMARY**

**Principal Risks**

**Specific Risks**

**Remedial Action**

1.  Immediate

2.  Short term (28 days)

3.  Medium term (6 months)

4.  Long term (over 12 months)

**PPE Maintenance Requirements**

**Information, Instruction, Training and Supervision Requirements**

**Date of next review**

● **FIG 3.6  Continued**

## DISPLAY SCREEN EQUIPMENT
## WORKSTATION RISK ASSESSMENT

| | Yes/No | Level of Risk | | |
|---|---|---|---|---|
| | | Low | Med | High |

**The Equipment**

**1. Display Screen**

Are the characters on the screen well defined and clearly formed, of adequate size and with adequate spacing between the characters and lines?

Is the image on the screen stable, with no flickering or other forms of instability?

Are the brightness and contrast between the characters and the background easily adjustable by the operator or user, and also easily adjustable to ambient conditions, e.g. lighting?

Does the screen swivel and tilt easily and freely to suit the needs of the operator?

Is it possible to use a separate base for the screen or an adjustable table?

Is the screen free of reflective glare and reflection liable to cause discomfort to the operator or user?

**2. Keyboard**

Is the keyboard tiltable and separate from the screen so as to allow the operator or user to find a comfortable working position avoiding fatigue in the arms or hands?

Is the space in front of the keyboard sufficient to provide support for the hands and arms of the operator or user?

Does the keyboard have a matt surface to avoid reflective glare?

Are the arrangement of the keyboard and the characteristics of the keys such as to facilitate the use of the keyboard?

Are the symbols on the keys adequately contrasted and legible from the design working position?

**3**

● **FIG 3.7 Display screen equipment workstation risk assessment**

| | Yes/No | Level of Risk | | |
|---|---|---|---|---|
| | | Low | Med | High |

### 3. Work Desk or Work Surface
Does the work desk or work surface have a sufficiently large, low reflectance surface and allow a flexible arrangement of the screen, keyboard, documents and related equipment?

Is the document holder stable and adjustable and positioned so as to minimise the need for uncomfortable head and eye movements?

Is there adequate space for operators or users to find a comfortable position?

### 4. Work Chair
Is the work chair stable and does it allow the operator or user easy freedom of movement and a comfortable position?

Is the seat adjustable in height?

Is the seat back adjustable in both height and tilt?

Is a footrest made available to any operator or user who requests one?

### Environment
### 1. Space requirements
Is the workstation dimensioned and designed so as to provide sufficient space for the operator or user to change position and vary movements?

### 2. Lighting
Does any room lighting or task lighting ensure satisfactory lighting conditions and an appropriate contrast between the screen and the background environment, taking into account the type of work and the vision requirements of the operator or user?

Are possible disturbing glare and reflections on the screen or other equipment prevented by co-ordinating workplace and workstation layout with the positioning and technical characteristics of artificial light sources?

● **FIG 3.7 Continued**

| | Yes/No | Level of Risk | | |
|---|---|---|---|---|
| | | Low | Med | High |

### 3.  Reflections and Glare

Is the workstation so designed that sources of light, such as windows and other openings, transparent or translucid walls, and brightly coloured fixtures or walls cause no direct glare and no distracting reflections on the screen?

Are windows fitted with a suitable system of adjustable covering to attenuate the daylight that falls on the workstation?

### 4.  Noise

Is the noise emitted  by equipment belonging to any workstation taken into account when a workstation is being equipped, with a view, in paticular, to ensuring that attention is not distracted and speech is not disturbed?

### 5.  Heat

Does the equipment belonging to any workstation produce excess heat which could cause discomfort to operators or users?

### 6.  Radiation

Is all radiation, with the exception of the visible part of the electromagnetic spectrum, reduced to negligible levels from the point of view of operators' or users' health and safety?

### 7.  Humidity

Are adequate levels of humidity established and maintained?

**3**

● **FIG 3.7  Continued**

| | Yes/No | Level of Risk | | |
|---|---|---|---|---|
| | | Low | Med | High |

**Interface Between Computer and Operator/User**

In the designing, selecting, commissioning and modifying of software, and in designing tasks using display screen equipment, does the employer take into account the following principles:–
(a) software must be suitable for the task;
(b) software must be easy to use and, where appropriate, adaptable to the level of knowledge or experience of the user; no quantitative or qualitative checking facility may be used without the knowledge of the operators or users;
(c) systems must provide feedback to operators or users on the performance of those systems;
(d) sytems must display information in a format and at a pace which are adapted to operators or users;
(e) the principles of software ergonomics must be applied, in particular to human data processing?

**Comments of Assessor**

Date _____ Signature _____

● **FIG 3.7 Continued**

## RISK ASSESSMENT SUMMARY

**Principal Risks**

**Specific Risks**

**Remedial Action**

1. Immediate

2. Short term (28 days)

3. Medium term (6 months)

4. Long term (over 12 months)

**Information, Instruction, Training and Supervision Requirements**

**Date of next review**

## MANUAL HANDLING OF LOADS

### EXAMPLE OF AN ASSESSMENT CHECKLIST

*Note:* This checklist may be copied freely. It will remind you of the main points to think about while you:
 – consider the risk of injury from manual handling operations
 – identify steps that can remove or reduce the risk
 – decide your priorities for action.

| SUMMARY OF ASSESSMENT | Overall priority for remedial action: Nil/Low/Med /High* |
|---|---|
| Operations covered by this assessment ...................... ............................................................................................... ............................................................................................... | Remedial action to be taken: ...................................... ............................................................................................... ............................................................................................... |
| Locations: ....................................................................... | Date by which action is to be taken: ......................... |
| Personnel involved: ...................................................... | Date of assessment: ................................................. |
| Date of assessment: ..................................................... | Assessor's name: .................Signature ..................... |

*circle as appropiate

### Section A – Preliminary:
Q1  Do the operations involve a significant risk of injury?                    Yes/No*
     If 'Yes' go to Q2. If 'No' the assessment need go no further.
     If in doubt answer 'Yes'. You may find the guidelines in Appendix 1 helpful.

Q2  Can the operations be avoided/mechanised/automated at reasonable cost?      Yes/No*
     If 'No' go to Q3. If 'Yes' proceed and then check that the result is satisfactory.

Q3  Are the operations clearly within the guidelines in Appendix 1?            Yes/No*
     If 'No' go to Section B. If 'Yes' you may go straight to Section C if you wish.

### Section C – Overall assessment of risk:
Q   What is your overall assessment of the risk of injury?          Insignificant/Low/Med/High*
     If not 'Insignificant' go to Section D. If 'Insignificant' the assessment need go no further.

### Section D – Remedial action:
Q   What remedial steps should be taken, in order of priority?

     i      ............................................................................................................................................

     ii     ............................................................................................................................................

     iii    ............................................................................................................................................

     iv     ............................................................................................................................................

     v      ............................................................................................................................................

### And finally:

 – complete the SUMMARY  above

 – compare it with your other manual handling assessments

 – decide your priorities for action

 – TAKE ACTION .....................AND CHECK THAT IT HAS THE DESIRED EFFECT

● **FIG 3.8  Factors to be considered when making an assessment of manual handling operations**

## Section B – More detailed assessment, where necessary

| Questions to consider:<br><br>(If the answer to a question is 'Yes' place a tick against it and then consider the level of risk) | | Level of risk:<br><br>(Tick as appropriate) | | | Possible remedial action<br><br>(Make rough notes in this column in preparation for completing Section D) |
|---|---|---|---|---|---|
| | Yes | Low | Med | High | |
| **The tasks** – do they involve: | | | | | |
| • holding loads away from trunk? | | | | | |
| • twisting? | | | | | |
| • stooping? | | | | | |
| • reaching upwards? | | | | | |
| • large vertical movement? | | | | | |
| • long carrying distances? | | | | | |
| • strenuous pushing or pulling? | | | | | |
| • unpredictable movement of loads? | | | | | |
| • repetitive handling? | | | | | |
| • insufficient rest or recovery? | | | | | |
| • a workrate imposed by a process? | | | | | |
| | | | | | |
| **The loads** – are they: | | | | | |
| • heavy? | | | | | |
| • bulky/unwieldy? | | | | | |
| • difficult to grasp? | | | | | |
| • unstable/unpredictable? | | | | | |
| • intrinsically harmful (eg sharp/hot?) | | | | | |
| | | | | | |
| **The working environment** – are there: | | | | | |
| • constraints on posture? | | | | | |
| • poor floors? | | | | | |
| • variations in levels? | | | | | |
| • hot/cold/humid conditions? | | | | | |
| • strong air movements? | | | | | |
| • poor lighting conditions? | | | | | |
| | | | | | |
| **Individual capability** – does the job: | | | | | |
| • require unusual capability? | | | | | |
| • hazard those with a health problem? | | | | | |
| • hazard those who are pregnant? | | | | | |
| • call for special information/training? | | | | | |
| | | | | | |
| **Other factors–**<br>Is movement or posture hindered by clothing or personal protective equipment? | | | | | |

Deciding the level of risk will inevitably call for judgement. The guidelines in Appendix 1 may provide a useful yardstick.

*When you have completed Section B go to Section C.*

**3**

# Safety Management Systems

The basis of successful safety management is the installation and maintenance of effective systems aimed principally at the prevention of accidents, ill-health and other forms of incident which result in loss to an organisation. Such systems should identify the standards to be maintained and the systems for monitoring and measuring performance in the achievement of these standards. Safety monitoring systems may be of an active nature, such as safety audits, safety inspections, health and safety reviews, or of a reactive nature, such as the reporting, recording and investigation of accidents, the use of various forms of accident rate and the examination of first aid treatments. All systems generate information on levels of performance, and effective systems of reporting, investigating, recording and analysing data are necessary to support them.

From an accident prevention viewpoint, by far the most important systems are those which monitor and measure performance and provide feedback to management and staff, i.e. active systems. While there are no specific rules or requirements relating to which form of health and safety monitoring should be carried out, different forms of monitoring can be used to cover specific situations or circumstances, and, principally, to provide an indication of the level of performance in an organisation, workplace, work activity or process. The various forms of active monitoring should include:

(a) indirect monitoring of performance standards where managers check the quantity and quality of monitoring activities undertaken by their subordinates

(b) procedures to monitor the achievement of objectives allocated to managers or sections, for example, by means of monthly or quarterly reports or returns

(c) the periodic examination of documents to check that standards relating to the promotion of the safety culture are complied with; for example, that suitable objectives have been established for each manager, that these are regularly reviewed, that all training needs have been assessed and recorded, and that the training needs are being met

(d) the systematic inspection of premises, plant and equipment by supervisors, maintenance staff or a joint team of management, safety representatives and other employees, to ensure the continued effective operation of hardware controls

(e) environmental monitoring and health surveillance to check on the effectiveness of health control measures and detect early signs of harm to health

(f) systematic direct observation of work and behaviour by first line supervisors to assess compliance with procedures, rules and systems – particularly when directly concerned with risk control.

## SAFETY MONITORING TECHNIQUES

Various forms of safety monitoring techniques are used in workplaces, with a view to measuring performance and identifying hazards. These include:

### *Safety sampling*

This technique is designed to measure by random sampling the accident potential in a specific work area or activity by identifying safety deficiencies or omissions. The area or activity is divided into sections and a trained observer appointed to each section. A prescribed route through the area is planned and observers follow their itinerary in the time allowed, perhaps 15 minutes During the sampling period they note specifically identified health and safety aspects on a sampling sheet, which incorporates a limited number of points to be awarded for each aspect, e.g. housekeeping – maximum 10 points; eye protection being worn – maximum 5 points; correct handling procedures being implemented – maximum 10 points. Typical aspects for observation and assessment would include observance of fire protection requirements, the state of hand tools, evidence of damage to machinery guards, the operation of specifically-designed safe systems of work and chemical handling procedures.

The staff making the observations should be trained in the technique and should have a broad knowledge of the procedures and processes being carried out. The results of the sample inspections are collated by a specific manager or health and safety adviser and presented in graphic form. This system, in effect, monitors the overall effectiveness of safety management in various areas of a workplace. An example of a safety sampling exercise form is shown at Fig. 4.1 (see p. 100).

### *Damage control*

Damage control and damage costing techniques emphasise the fact that non-injury accidents are as important as injury accidents. The elimination of non-injury accidents will, in many cases, remove the potential for other forms of accident.

For example, a pallet stack collapses and a falling pallet just misses an operator standing close by. No injury results and, therefore, no accident is recorded, but there may be damage to products, pallets, the building fabric, plant and fittings. However, the next time such a stack collapses, someone could be seriously injured or killed. This accident is recorded.

In these two situations, the correction of the cause of the first collapse, e.g. bad stacking, use of defective pallets, stacking too high or stacking on an uneven floor, would have prevented the injury following the second pallet stack collapse. Moreover, the independent assessment of costs incurred as a result of damage accidents, such as repairs to plant and equipment, replacement items, maintenance costs, structural repairs, labour costs, etc., will emphasise the need for safe systems of work and higher standards of supervision. In such cases, it may be appropriate to write these costs against an individual departmental manager's budget, which should have the effect of increasing his awareness of the cost of damage accidents. Managers must appreciate that evidence of damage to property, machinery and structural items is frequently an indication of poor safety performance.

Damage control techniques aim at providing a safe place of work, and call for the keenest observation and co-operation by all staff who see or experience a condition which may lead to an accident, e.g. a damaged machinery guard or a safety mechanism to a machine. Damage control relies heavily on the operation of a Hazard Reporting System covering damage and defects in plant, machinery, structural items, etc. An example of a Hazard Report Form is shown at Fig. 4.2 (see p. 101)

## Job safety analysis

Derived from task analysis or work study, job safety analysis, whether effected as part of work study or not, can do much to eliminate the hazards of a job. The analysis identifies every single operation in a job, examines the specific hazards and indicates remedial measures necessary. It involves the examination of a number of areas, including permit to work systems, influences on behaviour, the operator training required and the degree of supervision and control necessary. Job safety analysis can feature as part of a workplace risk assessment.

## Safety audits

A safety audit is a technique which submits each area of an organisation's health and safety activity to a systematic critical examination with the principal objective of minimising loss. Every component of the total working system is included, e.g. management policy and commitment, attitudes, training, features of processes, safe systems of work, personal protection needs, emergency procedures, etc. An audit, as in the field of accountancy, aims to disclose strengths and weaknesses in the main areas of vulnerability or risk. A specimen safety audit is shown at Fig. 4.3 (see p. 102).

Safety audits should be designed on the basis of past accident experience, existing hazards and the need to improve individual attitudes to health and safety performance. One of the problems with safety audits is that, while they cover a comprehensive range of issues in most cases, the people under-taking the audit only examine the issues raised in the audit documentation. On this basis, hazards can fail to be recognised due to the fact that the audit document does not incorporate any reference to them. In other words, a safe-ty audit is only as good as the individual who prepared the audit format. In order to be 100% effective, therefore, audit documents need regular updating and revision.

A team approach to the carrying out of safety audits is recommended, sooner than the completion of the audit by any specific individual. Following a safety audit, a formal report should be prepared, incorporating short-term, medium-term and long-term recommendations for action. Implementation of recommendations should be closely monitored.

## Safety surveys

A safety survey is a detailed examination of a number of critical areas of operation, e.g. process safety, manual handling operations, or an in-depth study of the whole health and safety operations of a workplace. A survey examines, for instance, health and safety management procedures, environ-mental working conditions, occupational health and hygiene arrangements, the wide field of safety and accident prevention, and the system for the pro-vision of information, instruction and training for staff and other persons affected by the organisation's operations.

The outcome of a safety survey is generally the production of a safety sur-vey report. In most cases such a report is purely of a critical nature indicat-ing, for instance, breaches of the law, unsafe conditions and working practices, inadequate management and administration of health and safety, deficiencies in machinery guarding, etc. The report is generally written on an Observation/Recommendation basis, recommendations being phased according to the degree of risk and the relative costs of eliminating or con-trolling these risks. The objective of a safety survey report is to present man-agement with a phased programme of health and safety improvement covering a five-year period. Following the safety survey, regular revisits are made to assess progress in the implementation of the report, progress reports being produced for senior management indicating the degree of progress and any other issues which may have arisen since the report was prepared.

## Safety inspections

This is a scheduled inspection of a workplace by a member of the management team and/or health and safety specialist. The inspection is generally broad in its application, covering plant maintenance standards, the operation of safe systems of work, cleaning and housekeeping levels, fire protection arrangements and other issues of specific significance in the workplace. This technique tends to be a more general examination of safety performance at a particular point in time, rather than the in-depth approach taken with a safety survey. The objectives of a safety inspection should be clearly defined and the outcome of the inspection should be a written report to individual managers with recommendations for action.

## Safety tours

A safety tour is an unscheduled examination of a work area undertaken by a selected group of staff, including the manager with direct responsibility for that area, members of a health and safety committee, supervisors, trade union safety representatives and selected operators. A safety tour can examine predetermined health and safety aspects, such as housekeeping levels, standards of machinery safety, the use of personal protective equipment and the operation of established safe systems of work. Safety tours should be related to and reinforce decisions made by local management or by the health and safety committee. For maximum effectiveness, it is essential that action following a safety tour is taken immediately.

## Project safety analysis

While existing risks may be identified and assessed by various forms of safety monitoring, project safety analysis, undertaken as a joint exercise by an engineering manager, architect, plant and equipment supplier and installer, health and safety specialist and other specialists, helps to ensure that account is taken of accumulated experience, knowledge of the technology and best practice in the initial design of projects. Project safety analysis should be undertaken at the design stage of all projects, both large and small. A typical analysis procedure is shown at Fig. 4.4 (see p. 108).

## Hazard and operability study (HAZOPS)

HAZOPS is a technique applied in the assessment of potential hazards from new installations and processes. It is a technique used extensively in the chemical industry and in chemical engineering applications, along with other techniques, such as Failure Modes and Effect Analysis, Fault Tree Analysis and Event Tree Analysis.

The technique is defined as 'the application of a formal critical examination to the process and engineering intentions of new facilities to assess the hazard potential from incorrect operation or malfunction of individual items of equipment, and the consequential effects on the facility as a whole'. It ensures that health and safety requirements are considered at the design stage of a project with a view to preventing the classic 'fire-fighting' exercises which commonly follow the completion of projects.

HAZOPS techniques use guide words such as 'too much', 'too little', which can be applied to the process parameters to generate 'what if?' questions, such as 'what if the operating temperature is greatly exceeded?' Remedial action can then be considered and implemented in the design stages of a particular project (see Fig. 4.5 – Safety monitoring of a typical project through various stages of its development, p. 113).

## Failure modes and effect analysis (FMEA)

This technique is based on identifying the possible failure modes of each component of a system and predicting the consequences of that failure. For example, if a control valve fails it could result in too much flow in the system, too much pressure, or the production of an undesired chemical reaction. As a result attention is paid to these consequences at the design stage of a project and in the preparation of planned maintenance systems.

## Fault tree analysis

Fundamentally, this begins with consideration of a chosen 'top event', such as a major fire or an explosion, and then assesses the combination of failures and conditions which could cause that event to arise. It is widely used in quantitative risk analysis, particularly where control over process controls is critical to meet safety standards.

## Event tree analysis

Similar to fault tree analysis, this works from a selected 'initiating event', such as a pressure control failure. It is, basically, a systematic representation of all the possible states of the processing system conditional to the specific initiating event and relevant for a certain type of outcome, e.g. a pollution incident or a major fire.

## Consequence analysis

Consequence analysis is a feature of a risk analysis which considers the physical effects of a particular process failure and the damage caused by these effects. It is undertaken to form an opinion on potentially serious hazardous outcomes of accidents and their possible consequences for people and the environment. Consequence analysis should act as a tool in the decision-making process in a safety study which incorporates the features described below:

(a)    description of the process system to be investigated

(b)    identification of the undesirable events

(c)  determination of the magnitude of the resulting physical effects

(d)  determination of the damage

(e)  estimation of the probability of the occurrence of calculated damage

(f)  assessment of risk against established criteria.

The outcome of consequence analysis is fourfold, thus:

(a)  for the chemical and process industries to obtain information about all known and unknown effects that are of importance when something goes wrong in the plant and also to get information on how to deal with possible catastrophic events

(b)  for the designing industries to obtain information on how to minimise the consequences of accidents

(c)  for the workers in the processing plant and people living in the immediate vicinity, to give them an understanding of their personal situation and the measures being taken to protect them

(d)  for the legislative authorities.

Consequence analysis is generally undertaken by professional technologists and chemists who are experienced in the actual problems of the technical system concerned.

The logical chain of consequence analysis is shown in Fig. 4.6 below.

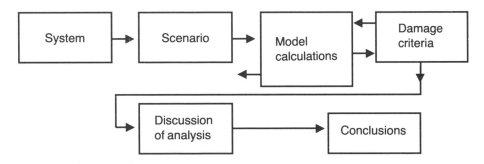

● **FIG 4.6 Logical chain of consequence analysis**

The first step in the chain is a description of the technical system to be investigated. In order to identify the undesirable events it is necessary to construct a scenario of possible incidents. The next stage is to carry out model calculations in which damage level criteria are taken into account. Following discussions by the assessment team, conclusions can be drawn as to possible consequences.

Feedback from model calculations to the scenario is included, since the linking of the outputs from the scenario to the inputs of models may cause difficulties. There is also another feedback, from damage criteria to model calculations, in case these criteria should be influenced by possible threshold values of the legislative authorities.

# ACCIDENT DATA

Various forms of accident data are collected by organisations and a number of standards indices are used, e.g. rates of accident incidence, frequency, severity and duration. Accident data are based on information compiled from accident reports. As such, they are a reactive form of safety monitoring and should not, of course, be used as the sole means of measuring safety performance. However, they do indicate trends in accident experience and provide feedback which can be incorporated in future accident prevention strategies. The following rates are used:

$$\text{Frequency rate} = \frac{\text{Total number of accidents}}{\text{Total number of man hours worked}} \times 1{,}000{,}000$$

$$\text{Incidence rate} = \frac{\text{Total number of accidents}}{\text{Average number of persons exposed}} \times 1000$$

$$\text{Severity rate} = \frac{\text{Total number of days lost}}{\text{Total man hours worked}} \times 1000$$

$$\text{Mean duration rate} = \frac{\text{Total number of days lost}}{\text{Total number of accidents}}$$

$$\text{Duration rate} = \frac{\text{Number of man hours worked}}{\text{Total number of accidents}}$$

However, there are problems with the use of accident data as a measure of performance. Studies by Amis and Booth (1991) questioned the significance and relevance of accident statistics as a measure of health and safety performance. Their conclusions were as follows:

1. They measure failure, not success.
2. They are difficult to use in staff appraisal.
3. They are subject to random fluctuations; there should not be enough accidents to carry out a statistical evaluation. Is safety fully controlled if, by chance, there are no accidents over a period?
4. They reflect the success or otherwise of safety measures taken some time ago. There is a time delay in judging the effectiveness of new measures.
5. They do not measure the incidence of occupational diseases where there is a prolonged latent period.
6. They measure injury severity, not necessarily the potential seriousness of the accident. Strictly, they do not even do this. Time off work as a result of injury may not correlate well with true injury severity. Data may be affected by the known variations in the propensity of people to take time off work for sickness in different parts of the country.
7. They may under-report (or over-report) injuries, and may vary as a result of subtle differences in reporting criteria.
8. They are particularly limited for assessing the future risk of high consequence, low probability accidents. (A fatal accident rate based on data from single fatalities may not be a good predictor of risk of multiple fatal emergencies.)

The crucial point is that counting numbers of accidents provides incomplete, untimely and possibly misleading answers to the questions –

1  Are we implementing fully our safety plan?

2  Is it the right plan?

Where management have not drawn up a safety plan, counting accidents is the only measure of safety performance available, apart, of course, from auditing compliance with statutory hardware requirements. This is a reason why safety management via accident rate comparison is attractive to the less competent and committed employer.

## ACCIDENT RATIOS

Studies of accidents resulting in major or lost time injury, minor injuries, property damage and 'near misses' over the last sixty years have produced a number of accident ratios. The principal objective of these studies was to establish a relationship between near misses and other categories of accident. The various accident ratios produced by Heinrich (1959), Bird (1966), the British Safety Council (1975) and the HSE (1993) are shown in Fig. 4.7 (see p. 114).

The philosophy behind these various accident ratios is that by investigating and taking action to prevent 'near misses', the more serious accidents and incidents would be prevented. This approach could be said to be far too simplistic, in that not all near misses involve risks which might have caused fatal or serious injury. However, these ratios are of great value in comparing categories of accidents in order to identify those categories which are most likely to cause serious harm.

### Accident investigation

The investigation of the direct and indirect causes of accidents is a reactive strategy in safety management. There are very good reasons, however, for the effective and thorough investigation of accidents:

(a)  on a purely humanitarian basis, no one wishes to see people killed or injured

(b)  the accident may have resulted from a breach of statute or regulations by the organisation, the accident victim, the manufacturers and/or suppliers of articles and substances used at work, or other persons, e.g. contractors, with the possibility of civil proceedings being instituted by the injured party against the employer and other persons

(c)  the injury may be reportable to the enforcing authority (HSE, local authority) under the Reporting of Injuries, Diseases and Dangerous Occurrences Regulations 1985 (RIDDOR)

(d)  the accident may result in lost production

(e)   from a management viewpoint, a serious accident, particularly a fatal one, can have a long-term detrimental effect on the morale of the work force and management/worker relations

(f)   there may be damage to plant and equipment, resulting in the need for repair or replacement, with possible delays in replacement

(g)   in most cases, there will be a need for immediate remedial action in order to prevent a recurrence of this accident.

Apart from the obvious direct and indirect losses associated with all forms of incident, not just those resulting in injury, there are legal reasons for investigating accidents to identify the direct and indirect causes and to produce strategies for preventing recurrences. Above all, the purpose of accident investigation is not to apportion blame or fault, although these may eventually emerge as a result of accident investigation.

### What accidents should be investigated?

Clearly there is a cause for investigating all accidents and, indeed, 'near misses'. A near miss is defined as 'an unplanned and unforeseeable event that could have resulted, but did not result in human injury, property damage or other form of loss'. However, it may be impracticable to investigate every accident, but the following factors should be considered in deciding which accidents should be investigated as a priority;

(a)   the type of accident, e.g. fall from a height, chemical handling, machinery-related

(b)   the form and severity of injury, or the potential for serious injury and/or damage

(c)   whether the accident indicates the continuation of a particular trend in the organisation's accident experience

(d)   the extent of involvement of articles and substances used at work, e.g. machinery, work equipment, hazardous substances, and ensuing damage or loss

(e)   the possibility of a breach of the law, e.g. HSWA

(f)   whether the injury or occurrence is, by law, notifiable and reportable to the enforcing authority

(g)   whether the accident should be reported to the insurance company as it could result in a claim being submitted.

### Practical accident investigation

In any serious incident situation, such as a fatal accident, or one resulting in major injury e.g. fractures, amputations, loss of an eye; or where there has been a scheduled dangerous occurrence, as listed in the Schedule to RIDDOR, such as the collapse of a crane, speed of action is essential. This is particularly the case when it comes to interviewing injured persons and witnesses. The following procedure is recommended:

1.  Establish the facts as quickly and completely as possible with regard to:
    (a)  the general environment
    (b)  the particular plant, machinery, practice or system of work involved
    (c)  the sequence of events leading to the accident.
2.  Use an instant camera to take photographs of the accident scene prior to any clearing-up that may follow the accident.
3.  Draw sketches and take measurements with a view to producing a scale drawing of the events leading up to the accident.
4.  List the names of all witnesses, i.e. those who saw, heard, felt or smelt anything; interviewing them thoroughly in the presence of a third party if necessary, and take full statements. In certain cases, it may be necessary to formally caution witnesses prior to their making a statement. Do not prompt or lead witnesses.
5.  Evaluate the facts, and individual witnesses' versions of them, as to accuracy, reliability and relevance.
6.  Endeavour to arrive at conclusions as to the causes of the accident on the basis of the relevant facts.
7.  Examine closely any contradictory evidence. Never dismiss a fact that does not fit in with the rest. Find out more.
8.  Learn fully about the system of work involved. Every accident occurs within the context of a work system. Consider the people involved in terms of their ages, training, experience and level of supervision, and the nature of the work, e.g. routine, sporadic or incidental.
9.  In certain cases it may be necessary for plant and equipment, such as lifting appliances, machinery and vehicles to be examined by a specialist, e.g. a consultant engineer.
10. Produce a report for the responsible manager emphasising the causes and remedies to prevent a recurrence, including any changes necessary.
11. In complex and serious cases, consider the establishment of an investigating committee comprising managers, supervisors, technical specialists and trade union safety representatives.

There are other persons who may also wish to investigate this accident, including:

(a)  trade union safety representatives, should a member of their trade union be injured
(b)  insurance company liability surveyor, in the event of a claim being submitted by the organisation
(c)  legal representative of the injured party to establish the cause of the accident and whether there has been a breach of statutory duty or negligence on the part of the employer
(d)  officers of the enforcing authority, to establish whether there has been a breach of current health and safety legislation which may require

action, such as the service of a Prohibition or Improvement Notice, or prosecution.

It is essential, therefore, that any accident report produced is accurate, comprehensible and identifies the causes of the accident, both direct and indirect.

## The outcome of accident investigation

Whether the investigation of an accident is undertaken by an individual, e.g. health and safety specialist, or by a special committee, it is necessary, once the causes have been identified, to submit recommendations to management with a view to preventing a recurrence. The organisation of feedback on the causes of accidents is crucial in large organisations, especially those who operate more than one site or premises. An effective investigation should result in one or more of the following recommendations being made:

(a) the issuing of specific instructions by management covering, for instance, systems of work, the need for more effective guarding of machinery or safe manual handling procedures

(b) the establishment of a working party or committee to undertake further investigation, perhaps in conjunction with members of the safety committee and/or safety representatives

(c) the preparation and issue of specific codes of practice or guidance notes dealing with the procedures necessary to minimise a particular risk, e.g. the use of a permit to work system

(d) the identification of specific training needs for groups of individuals e.g. managers, foremen, supervisors, machinery operators, drivers, and the implementation of a training programme designed to meet these needs

(e) the formal analysis of the job or system in question, perhaps using job safety analysis techniques, to identify skill, knowledge and safety components of the job

(f) identification of the need for further information relating to articles and substances used at work, e.g. work equipment, chemical substances

(g) identification of the need for better environmental control, e.g. noise reduction at source or improved lighting

(h) general employee involvement in health and safety issues, e.g. the establishment of a health and safety committee

(i) identification of the specific responsibilities of groups with regard to safe working practices.

Above all, a system of monitoring should be implemented to ensure that the lessons which have been learned from the accident are put into practice or incorporated in future systems of work, and that procedures and operating systems have been produced for all grades of staff.

# THE COST OF ACCIDENTS AT WORK

The term 'accident' refers to personal injury and 'loss'. Given that a minimum level of profitability is essential for a business, the prevention of losses is a key factor. All accidents, whether they result in personal injury, property and plant damage and/or interruption of the business activity, represent losses to an organisation. Experience indicates that managers frequently complain about 'the cost of safety'. What they never consider are the losses to the organisation as a result of accidents!

## The direct costs

Direct costs, sometimes referred to as 'insured costs', are largely concerned with an organisation's liabilities as an employer and occupier of premises. Direct costs are covered by premiums paid to an insurance company to provide cover against claims made by an injured party. Premiums paid are determined, to some extent, by the claims history of the organisation and the risks involved in the business activities.

Other direct costs of accidents are claims by insured persons and users of products manufactured by the company, which are settled either in or out of a court, together with fines imposed by the courts for breaches of health and safety legislation. Substantial legal defence costs may also be incorporated in this category.

## The indirect costs

These are many and varied and are often difficult to predict. Some indirect costs may be included, and thus hidden, in other costs, e.g. labour costs, production and administration costs. It is common for indirect costs to be ignored owing to the difficulty experience in separating them from other costs. Some indirect costs, however, are simply to quantify. These are outlined below.

1. *Treatment costs* First aid, transport to hospital, hospital charges, other costs, e.g. local doctor, consultants.
2. *Lost time costs* Lost time of injured person, supervisor, first aiders and others involved.
3. *Production costs* Lost production, extra staff payments to meet production targets, damage to plant, vehicles, raw materials and finished products.
4. *Training and supervision costs* Training for replacement and existing labour force, extra supervision costs.
5. *Investigation costs* A number of people may be involved in the investigation of an accident or incident – management, safety representatives, safety adviser, etc. Investigation costs should be based on the total man-hours involved by all concerned.
6. *Miscellaneous costs* Ex-gratia payments, perhaps to a widow, replacement

costs of personal items belonging to the injured person, incidental costs incurred by witnesses, etc.

7. *Costs to the State* These are difficult to quantify but can be enormous if surgery and a prolonged stay in hospital results from the accident.
8. *Costs to the injured person* Loss of earnings, loss of total earning capacity, legal costs in pursuing injury claim, possible legal costs in defending a prosecution for 'unsafe behaviour' at work.

### Accident Costing

All accidents should be costed using a standard accident costing form (see example, Fig. 4.8, p. 115) .

# EMERGENCY PROCEDURES

The risk assessment required under Regulation 3 of the MHSWR should identify the significant risks arising out of work. These could include, for instance, the potential for a major escalating fire, explosion, building collapse, pollution incident, bomb threat and some of the scheduled dangerous occurrences listed in RIDDOR, e.g. the explosion, collapse or bursting of any closed pressure vessel. All these events could result in a major incident, which can be defined as one that may:

(a)   affect several departments within an undertaking

(b)   endanger the surrounding communities

(c)   be classed as a dangerous occurrence under RIDDOR

(d)   result in adverse publicity for the organisation with ensuing loss of public confidence and market place image.

Therefore the question must be asked – 'What are the worst possible types of incident that could arise from the process or undertaking?' Once these major risks, which could result in serious and imminent danger, have been identified, a formal emergency procedure must be produced.

The ACOP, which should be read in conjunction with RIDDOR, raises a number of important points with regard to the establishment of emergency procedures.

1. The aim must be to set out clear guidance on when employees and others at work should stop work and how they should move to a place of safety.
2. The risk assessment should identify the foreseeable events that need to be covered by these procedures.
3. Many workplaces or work activities will pose additional risks. All employers should consider carefully in their risk assessment whether such additional risks might arise.

4.  The procedures may need to take account of responsibilities of specific employees, e.g. in the shutting-down of plant.

5.  The procedures should set out the role, identity and responsibilities of the competent persons nominated to implement the detailed actions.

6.  Where specific emergency situations are covered by particular Regulations, procedures should reflect any requirements laid on them by these Regulations.

7.  The procedure should cater for the fact that emergency events can occur and develop rapidly, thus requiring employees to act without waiting for further guidance.

8.  Emergency procedures should normally be written down, clearly setting out the limits of action to be taken by all employees.

9.  Work should not be resumed after an emergency if a serious danger remains. Consult the emergency authorities if in doubt.

10. In shared workplaces, separate emergency procedures should take account of others in the workplace and, as far as is appropriate, should be co-ordinated.

## Establishing the emergency procedure

The risk assessment undertaken to comply with the MHSWR should identify those highly significant risks where an emergency procedure is essential. A properly conceived emergency procedure will take account of four phases or stages of an emergency.

### Phase 1 – Preliminary action

This should cover:

(a) the preparation of a plan, tailored to meet the special requirements of the site, products and surroundings, including –
   (i)   a list of all key telephone numbers
   (ii)  the system for the provision of emergency lighting, e.g. hand lamps and torches
   (iii) the designation of exit routes
   (iv)  a plan of the site layout identifying hydrant points and the location of shut-off valves to energy supplies, e.g. gas
   (v)   notes on specific hazards on site for use by the emergency services

(b) the familiarisation through training of every employee with the details of the plan, including the position of essential equipment

(c) the training of personnel involved, in particular, key personnel

(d) the initiation of a programme of inspection of potentially hazardous areas, testing of warning systems and evacuation procedures

(e) stipulating specific periods at which the plan is to be re-examined and updated

### Phase 2 – Action when emergency is imminent

There may be a warning of the emergency, in which case this period should be used to assemble key personnel, to review the standing arrangements in order to consider whether changes are necessary, to give advance warning to the external authorities, and to test all systems connected with the emergency procedure.

### Phase 3 – Action during emergency

If Phase 1 has been properly executed, and Phase 2 where applicable, Phase 3 proceeds according to plan. However, it is likely that unexpected variations in a predicted emergency will take place. The decision-making personnel, selected beforehand for this purpose, must be able to make precise and rapid judgements and see that the proper action follows the decisions made.

### Phase 4 – Ending the emergency

There must be a procedure for clearing plant, systems and specific areas safe, together with an early reoccupation of buildings where possible.

## Implementing the emergency procedure

The following matters must be taken into account when implementing an emergency procedure.

### Liaison with external authorities and other companies

The closest contact must be maintained with the fire, police, ambulance and health authorities, together with the Health and Safety Executive and local authority. A mutual aid scheme involving neighbouring premises is best undertaken at this stage. A major emergency may involve a failure in the supply of gas, electricity, water and/or telephone communications. Discussions with the appropriate authority will help to determine priorities in re-establishing supply.

### Emergency controller

A senior manager, with thorough knowledge of all processes and their associated hazards, should be nominated emergency controller, and a deputy appointed to cover absence, however brief this may be. Out of normal working hours, the senior member of management should take initial control until the emergency controller arrives.

### Emergency control centre

A sound communication system is essential if a major emergency is to be handled effectively. A control centre should be established and equipped with means of receiving information from the forward control and assembly points, transmitting calls for assistance to external authorities, calling in essential personnel and transmitting information and instructions to personnel within the premises. Alternative means of communication must be available in the event of the main system being rendered inoperative, e.g. field telephones. A fall-back control centre may be necessary in certain situations such as a rapidly-escalating fire.

### Initiating the procedure
The special procedure for handling major emergencies must only be initiated when such an emergency is known to exist. A limited number of designated senior managers should be assigned the responsibility of deciding if a major emergency exists or is imminent. Only these persons should have authority to implement the procedure.

### Notification to local authorities
Notification can be achieved by a predetermined short message, transmitted by an emergency line or by the British Telecom lines. The warning message should mention routes to the premises which may become impassable. Alternative routes can then be used.

### Call out of key personnel
A list of key personnel required in the event of a major emergency should be drawn up, together with the internal and home telephone numbers and addresses. The list should be available in control centres and constantly updated.

### Immediate action on site
An emergency would be dealt with by action by supervisors and operators designed to close down and make safe those parts which are affected or likely to be affected (danger areas). Preservation of human life and the protection of property are of prime importance, and injured persons should be conveyed to hospital with the least possible delay. This may require temporary facilities at points in a safe area accessible to ambulances.

### Evacuation
Complete evacuation of non-essential personnel immediately the alarm is sounded is usually advisable, though it may not be necessary or advisable in large workplaces. In either situation, however, an evacuation alarm system should be installed and made known to all employees, for the purpose of evacuating the premises. Evacuation should be immediately followed by a roll call at a prescribed assembly point.

### Access to records
Because relatives of injured and/or deceased employees will have to be informed by the police, each control centres should keep a list of names and addresses of all employees.

### Public relations
As a major incident will attract the attention of the media, it is essential to

make arrangements for official releases of information to the press and other news services. This is best achieved through a designated public relations officer. Other employees should be instructed not to release information, but to refer any enquiries to this officer, who should keep a record of any media enquiries dealt with during the emergency.

### Catering and temporary shelter

Emergency teams will need refreshment and temporary shelter if the incident is of long duration. Where facilities on the premises cannot be used it may be possible for the local authority or neighbouring companies to provide facilities.

### Contingency arrangements

A contingency plan should be drawn up covering arrangements for repairs to buildings, drying out and temporary waterproofing, replacement of raw materials, alternative storage and transport arrangements.

### Training

Training is an important feature of any emergency procedure. Training exercises should include the participation of external services, such as the fire brigade, ambulance service and police. Where mutual aid schemes exist with neighbouring organisations, all possible participants should take part in any form of training exercise.

### Statement of health and safety policy

Familiarisation of all staff with the procedure, together with training exercises at regular intervals, will help reduce the risk of fatal and serious injuries following an emergency. For this reason the company emergency procedure should be linked to the Statement of Health and Safety Policy, perhaps as a specific Code of Practice referred to in the Statement.

## SAFETY SAMPLING EXERCISE

| | | Area A | Area B | Area C | Area D | Area E | Area F |
|---|---|---|---|---|---|---|---|
| 1. | Housekeeping/cleaning (Max. 10) | | | | | | |
| 2. | Personal protection (Max. 10) | | | | | | |
| 3. | Machinery (Max. 10) | | | | | | |
| 4. | Chemical storage (Max. 5) | | | | | | |
| 5. | Chemical handling (Max. 5) | | | | | | |
| 6. | Manual handling (Max. 5) | | | | | | |
| 7. | Fire protection (Max. 10) | | | | | | |
| 8. | Structural hazards (Max. 10) | | | | | | |
| 9. | Internal transport. e.g. FLTs (Max. 5) | | | | | | |
| 10. | Access equipment (Max. 5) | | | | | | |
| 11. | First Aid boxes (Max. 5) | | | | | | |
| 12. | Hand tools (Max. 10) | | | | | | |
| 13. | Internal storage (racking systems) (Max. 5) | | | | | | |
| 14. | Structural safety (Max. 10) | | | | | | |
| 15. | Temperature (Max. 5) | | | | | | |
| 16. | Lighting (Max. 5) | | | | | | |
| 17. | Ventiliation (Max. 5) | | | | | | |
| 18. | Noise (Max. 5) | | | | | | |
| 19. | Dust and fumes (Max. 5) | | | | | | |
| 20. | Welfare amenities (Max. 10) | | | | | | |
| | TOTALS– (Max. 140) | | | | | | |

● **FIG 4.1 Safety sampling exercise (see text p. 83)**

---

## HAZARD REPORT

1. **Report** (to be completed by persons identifying hazard).

Date ..............................Time ...........................Department...........................

Reported to:- Verbal ..........................Written........................................(Names)

Description of Hazard (including location, plant, machinery, etc.) ......................

...........................................................................................................

...........................................................................................................

Signature...........................................Position ......................................

---

2. **Action** (to be completed by Departmental Manager/Supervisor)

   \*Hazard verified YES/NO   Date ...........................Time ...............................

Remedial action (including changes in system of work)........................................

...........................................................................................................

Action to be taken by:- Name .........................Signature................................

+\* Priority rating: 1 2 3 4 5 Estimated Cost...................................................

Completion:Date...........................................Time ...............................

Interim Precautions ...............................................................................

...........................................................................................................

Signature ...............................................................................(Dept. Manager)

---

3. **Financial Approval** (to be completed by  Manager or his Assistant where cost exceeds departmental authority).
The expenditure necessary to complete the above work is approved.

Signature..............................(Manager/Assistant Manager)  Date .....................

---

4. **Completion** The remedial action described above is complete.

Actual cost .............................................................................................

Date...........................................Date ......................................................

Signatures.....................(Persons completing work) ......................(Dept. Head)

---

5. **Safety Officer's Check**  I have checked completion of the above and confirm that the hazards has been eliminated.

Signature.......................(Safety Officer)    Date/Time ......................................

---

\*Delete as appropriate
+Priority Ratings 1 (immediate) 2 (48 hours) 3 (1 week) 4 (1 month) 5 (3 months)

● **FIG 4.2  Hazard report (see p. 84)**

<h2 style="text-align:center;">SAFETY AUDIT</h2>

| | Yes/No |
|---|---|

**Documentation**

1.  Are management aware of all health and safety legislation applying to their workplace?

    Is this legislation available to management and employees?

2.  Have all Approved Codes of Practice, HSE Guidance Notes and internal codes of practice been studied by management with a view to ensuring compliance?

3.  Does the existing Statement of Health and Safety Policy meet current conditions in the workplace?

    Is there a named manager with overall responsibility for health and safety?

    Are the 'organisation and arrangements' to implement the Health and Safety Policy still adequate?

    Have the hazards and precautions necessary on the part of staff and other persons been identified and recorded?

    Are individual responsibilities for health and safety clearly detailed in the Statement?

4.  Do all job descriptions adequately describe individual health and safety responsibilities and accountabilities?

5.  Do written safe systems of work exist for all potentially hazardous operations?

    Is Permit to Work documentation available?

6.  Has a suitable and sufficient assessment of the risks to staff and other persons been made, recorded and brought to the attention of staff and other persons?

    Have other risk assessments in regard to:

    (a)  substances hazardous to health

    (b)  risks to hearing

    (c)  work equipment

    (d)  personal protective equipment

    (e)  manual handling operations

● **FIG 4.3 Safety audit document (see text p. 84)**

been made, recorded and brought to the attention of staff and other persons?

7. Is the fire certificate available and up to date?

   Is there a record of inspections of the means of escape in the event of fire, fire appliances, fire alarms, warning notices, fire and smoke detection equipment?

8. Is there a record of inspections and maintenance of work equipment, including guards and safety devices?

   Are all examination and test certificates available, e.g. lifting appliances and pressure systems?

9. Are all necessary licences available, e.g. to store petroleum spirit?

10 Are workplace health and safety rules and procedures available, promoted and enforced?

   Have these rules and procedures been documented in a way in which is comprehensible to staff and others, e.g. Health and Safety Handbook?

   Are disciplinary procedures for unsafe behaviour clearly documented and known to staff and other persons?

11 Is a formally written emergency procedure available?

12 Is documentation available for the recording of injuries, near misses, damage only accidents, diseases and dangerous occurrences?

13 Are health and safety training records maintained?

14 Are there documented procedures for regulating the activities of contractors, visitors and other persons working on the site?

15 Is hazard reporting documentation available to staff and other persons?

16. Is there a documented planned maintenance system?

17. Are there written cleaning schedules?

**Health and Safety systems**

1. Have competent persons been appointed to:

   (a) co-ordinate health and safety measures

Continued

(b)  implement the emergency procedure?

Have these persons been adequately trained on the basis of identified and assessed risks?

Are the role, function, responsibility and accountability of competent persons clearly identified?

2.  Are there arrangements for specific forms of safety monitoring, e.g. safety inspections, safety sampling?

Is a system in operation for measuring and monitoring individual management performance on health and safety issues?

3.  Are systems established for the formal investigation of accidents, ill-health, near misses and dangerous occurrences?

Do investigation procedures produce results which can be used to prevent future incidents?

Are the causes of accidents, ill-health, near misses and dangerous occurrences analysed in terms of failure of established safe systems of work?

4.  Is a hazard reporting system in operation?

5.  Is a system for controlling damage to structural items, machinery, vehicles, etc. in operation?

6.  Is the system for joint consultation with trade union safety representatives and staff effective?

Are the role, constitution and objectives of the Health and Safety Committee clearly identified?

Are the procedures for appointing or electing Committee members and trade union safety representatives clearly identified?

Are the available facilities, including training arrangements, known to committee members and trade union safety representatives?

7.  Are the capabilities of employees as regards health and safety taken into account when entrusting them with tasks?

8.  Is the provision of first aid arrangements adequate?

Are first aid personnel adequately trained and retrained?

9.  Are the procedures covering sickness absence known to staff?

Is there a procedure for controlling sickness absence?

Are managers aware of the current sickness absence rate?

Continued

10. Do current arrangements ensure that health and safety implications are considered at the design stage of projects?

11. Is there a formally-established annual health and safety budget?

**Prevention and control procedures**

1. Are formal inspections of machinery, plant, hand tools, access equipment, electrical equipment, storage equipment, warning systems, first-aid boxes, resuscitation equipment, welfare amenity areas, etc. undertaken?

   Are machinery guards and safety devices examined on a regular basis?

2. Is a Permit to Work system operated where there is a high degree of foreseeable risk?

3. Are fire and emergency procedures practised on a regular basis?

   Where specific fire hazards have been identified, are they catered for in the current fire protection arrangements?

   Are all items of fire protection equipment and alarms tested, examined and maintained on a regular basis?

   Are all fire exits and escape routes marked, kept free from obstruction and operational?

   Are all fire appliances correctly labelled, sited and maintained?

4. Is a planned maintenance system in operation?

5. Are the requirements of cleaning schedules monitored?

   Is housekeeping of a high standard, e.g. material storage, waste disposal, removal of spillages?

   Are all gangways, stairways, fire exits, access and egress points to the workplace maintained and kept clear?

6. Is environmental monitoring of temperature, lighting, ventilation, humidity, radiation, noise and vibration undertaken on a regular basis?

7. Is health surveillance of persons exposed to assessed health risks undertaken on a regular basis?

8. Is monitoring of personal exposure to assessed health risks undertaken on a regular basis?

Continued

4

9.  Are local exhaust ventilation systems examined, tested and maintained on a regular basis?

10. Are arrangements for the storage and handling of substances hazardous to health adequate?

    Are all substances hazardous to health identified and correctly labelled, including transfer containers?

11. Is the appropriate personal protective equipment available?

    Is the personal protective equipment worn or used by staff consistently when exposed to risks?

    Are storage facilities provided for items of personal protective equipment?

12. Are welfare amenity provisions, i.e. sanitation, hand washing, showers and clothing storage arrangements adequate?

    Do welfare amenity provisions promote appropriate levels of personal hygiene?

**Information, instruction, training and supervision**

1.  Is the information provided by manufacturers and suppliers of articles and substances for use at work adequate?
    Do employees and other persons have access to this information?

2.  Is the means of promoting health and safety adequate?
    Is effective use made of safety propaganda, e.g. posters?

3.  Do current safety signs meet the requirements of the Safety Signs Regulations 1980?
    Are safety signs adequate in terms of the assessed risks?

4.  Are fire instructions prominently displayed?

5.  Are hazard warning systems adequate?

6.  Are the individual training needs of staff and other persons assessed on a regular basis?

7.  Is staff health and safety training undertaken:-

    (a) at the induction stage

    (b) on their being exposed to new or increased risks because of:
        (i) transfer or change in responsibilities

Continued

    (ii)   the introduction of new work equipment or a
           change respecting existing work equipment

    (iii)  the introduction of new technology

    (iv)  the introduction of a new system of work or
           change in an existing system of work.

Is the above training:

(a)   repeated periodically

(b)   adapted to take account of new or changed risks

(c)   carried out during working hours?

8.   Is specific training carried out regularly for first-aid staff, fork lift truck drivers, crane drivers and others exposed to specific risks?

    Are selected staff trained in the correct use of fire appliances?

**Final question**

Are you satisfied that your organisation is as safe and healthy as you can reasonably make it, or that you know what action must be taken to achieve that state?

## PROJECT SAFETY ANALYSIS

### DANGEROUS SUBSTANCES

1. Lists substances which are:

   (a) flammable;

   (b) explosive;

   (c) corrosive;

   (d) toxic (state effects); and/or

   (e) have other special hazards (state hazards)

2. State whether:

   (a) raw material;

   (b) intermediate product;

   (c) final product or by-product; or

   (d) waste material

3. List the points where such substances are encountered in the process, by process description, equipment and cross reference to operating process manual.

4. List the significant physical properties, including:

   (a) incompatibility with other chemicals;

   (b) chemical reaction rates;

   (c) conditions of instability; and

   (d) other pertinent properties.

5. Show source of data and list available information sources on critical points.

### Process hazards

1. List maximum operating pressures under both normal and abnormal operating conditions.

2. State the form of pressure relief provided.
   State the location of pressure relief provided.
   State the condition of the relief devices.

3. State the date when the relief devices were last tested.
   State whether personnel are exposed to risk or injury on discharge of emergency relief devices.

● **FIG 4.4 Project safety analysis (see text p. 86)**

4. List the maximum permissible operating temperatures and sources of heat.

   Identify the overtemperature controls that are provided for abnormal operating conditions.

   State the protection provided to hot surfaces to protect personnel from burns.

5. State the dangers that may be present if process reaction conditions are deviated from in the manner below, the protection procedures necessary:

   (a)  abnormal temperature;

   (b)  abnormal reaction times;

   (c)  instrument failure;

   (d)  adding materials at the wrong stage;

   (e)  the wrong material added;

   (f)  material flow stoppage;

   (g)  equipment leaks, both out of the process and into the process;

   (h)  agitation failure;

   (i)  loss on inert gas blanket;

   (j)  error in valve or switch operation;

   (k)  blocked relief line;

   (l)  failure of relief device; and

   (m)  material spillage on floor or dispersal to air.

## Waste disposal

1. List gaseous stack effluents and concentrations, together with smoke characteristics.

2. State the approved height of the stacks.

   State whether scrubbers, electrostatic or centrifugal removal of stack effluents is needed.

3. State the direction of prevailing winds as they relate to exposed areas.

4. State the effluents which are run through waste disposal, from any point in the process, and the method of transfer.

   State their pH, relative toxicity, flammability and miscibility with water.

5. State whether waste chemicals can react with other waste chemicals in waste disposal systems and create hazards/difficulties.

6. State the procedure for preventing flammable liquids from reaching sewers.

7. List any special hazardous solid waste products, and the procedure for handling same, e.g. asbestos.

Continued

**Ventilation**

1. State the frequency of air changes required.

   State the frequency of checking ventilation equipment on a regular basis and the responsibility for same.

2. State whether exhaust ventilation is required at specific processes.

   List the air flow rates to be achieved and the responsibility for checking same.

3. State the risk of the ventilation intakes recirculating contaminants.

**Piping systems**

1. Confirm that piping systems are adequately supported with permanent hangers.

2. Confirm that pipework is of proper material and scheduled thickness for service.

3. Confirm that pressure tests for critical services and processes are scheduled on a regular basis.

4. Confirm that any bumping and/or tripping hazards are protected.

5. Confirm that safe access is provided to all valves.

**Electrical equipment**

1. Confirm that all hazardous locations are classified.

2. Confirm that all electrical equipment complies with the above including:

   (a) lighting;

   (b) wiring and switches;

   (c) motors;

   (d) instrumentation; and

   (e) intercoms, telephones, clocks etc.

3. Confirm all earthing meets the required standard.

4. Identify the person responsible for checking the above and frequency of checking.

Continued

### Access

1. Confirm two routes of access are provided to all occupied parts of buildings.

### Working platforms

1. Confirm safety rails with toe boards are provided on all platforms over 1m. high and all occupied enclosed portions of roofs.

2. Confirm safe means of access is provided to all working platforms.

### Machinery

1. Confirm all moving parts of plant and equipment are properly safeguarded to BS5304 standards.

2. Confirm there is a system for reporting and recording machinery and plant defects.

### Fire protection

1. Confirm all fire doors are checked on a regular basis.

2. Confirm exposed steel supporting major items of equipment is fire-proofed.

3. Confirm whether automatic or manually-operated sprinkler protection system is installed.

4. Confirm whether special fire extinguishing equipment is provided.

5. State the number, type and location of fire appliances and the system for ensuring regular servicing of such appliances.

6. Identify the location of fire hydrants and hose reels.

7. Identify the locations of fire alarm boxes and building evacuation alarms.

8. State whether any flammable substances are handled in the open.

9. Specify the amount stored and the location of flammable substances in operating buildings.

10. Specify the amount stored, location and the system for protection of flammable liquids and gases stored outdoors.

### Personnel, equipment and facilities

1. Specify the type of protective overclothing provided.

Continued

2.  Confirm that safety boots/shoes are provided.

3.  Confirm that gloves/gauntlets are provided for certain jobs.

4.  Confirm that the correct type of eye protection is provided where there is a risk of eye injury.

5.  Confirm that safety helmets/bump caps are provided for all operators.

6.  Confirm that respirator stations are adequately identified and maintained.

7.  State the frequency of overall changing, particularly where there is very heavy soiling.

8.  Confirm that amenity block of W.C.s, urinals, wash basins, showers, hot and cold water, and a separate mess room is provided.

9.  Confirm that a first aid facility is provided, together with trained first aid staff.

**Training**

1.  Confirm that operating manuals are available and have been provided.

2.  Confirm that the health and safety components of routine training for operators has been identified.

3.  Confirm there is a training schedule, and that staff are adequately trained.

4.  Confirm that the effectiveness of training is monitored.

5.  Confirm that occupational health practices are adequately covered in training.

6.  Confirm that safety rules have been written, published and are enforced.

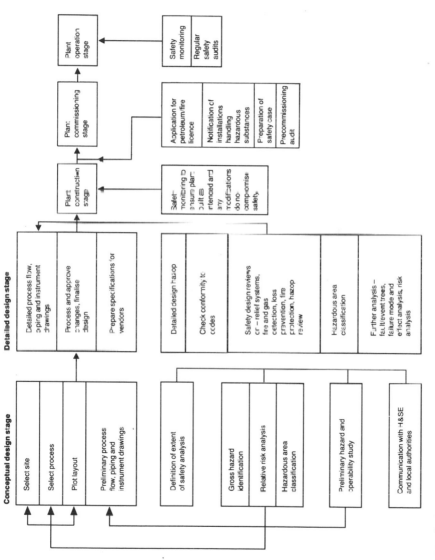

● **FIG 4.5  Safety monitoring of a typical project through various stages of its development (see text p. 87)**

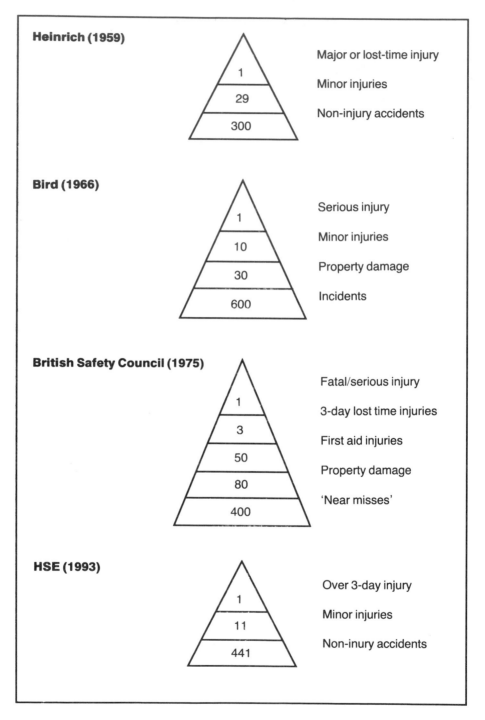

**Heinrich (1959)**

1 — Major or lost-time injury

29 — Minor injuries

300 — Non-injury accidents

**Bird (1966)**

1 — Serious injury

10 — Minor injuries

30 — Property damage

600 — Incidents

**British Safety Council (1975)**

1 — Fatal/serious injury

3 — 3-day lost time injuries

50 — First aid injuries

80 — Property damage

400 — 'Near misses'

**HSE (1993)**

1 — Over 3-day injury

11 — Minor injuries

441 — Non-inury accidents

● **FIG 4.7  Accident ratios (see text p. 90)**

---

## ACCIDENT COSTS ASSESSMENT

Date _____ Time _____ Place of accident _____

Details of accident

**Injured Person**
Name in full
Address
Occupation                          Length of service
Injury details

| **Accident costs** | £ | p |
|---|---|---|
| Direct costs | | |
| 1.  % occupier's liability premium | | |
| 2.  % increased premiums payable | | |
| 3.  Claims | | |
| 4.  Fines and damages awarded in court | | |
| 5.  Court and legal representation cost | | |
| **Indirect costs** | | |
| 6   Treatment     First aid | | |
|                  Transport | | |
|                  Hospital | | |
|                  Others | | |
| 7.  Lost time     Injured person | | |
|                  Management | | |
|                  Supervisors | | |
|                  First aiders | | |
|                  Others | | |
| 8.  Production/Lost production | | |
|                  Overtime payments | | |
|                  Damage to plant, vehicles, etc. | | |
|                  Training/supervision replacement labour | | |
| 9.  Investigation     Management | | |
|                  Safety adviser | | |
|                  Others e.g. safety representatives | | |
|                  Liaison with enforcement authority | | |
| 10. Other costs     Replacement of personal items – injured person others | | |
|                  Other miscellanous costs | | |
| **Total costs** | | |

4

---

● **FIG 4.8  Accident costing form (see text p. 95)**

# Occupational Health Management

Despite the fact that legislation is designed to cover both occupational health and occupational safety (hence the Health and Safety at Work etc. Act 1974), traditionally health and safety management has concentrated on the management of safety features, with little or no attention to the health of people at work. Regulations, such as the COSHH Regulations, the Noise at Work Regulations 1989 and the MHSWR 1992, have, however, greatly reinforced and extended the general duties of employers under the HSWA to pay attention to the health of employees, with the legal duty to undertake health risk assessments and to provide health surveillance where specific risks are identified. In many cases, this has meant the employment by organisations of occupational health practitioners or the use of external occupational health services.

Occupational health is defined as a branch of preventive medicine which examines, firstly, the relationship of work to health and, secondly, the effect of work upon the worker. Occupational health practitioners include occupational health nurses and occupational physicians, together with local general practitioners in certain cases. Occupational hygienists, health and safety practitioners and the trained first-aider also have a specific contribution to make in the provision and maintenance of healthy conditions at work. In certain cases, groups of occupational health practitioners are combined as an occupational health service, operating either within a specific organisation or locally, providing a service to subscribing companies in the area. One of the principal functions of an occupational health service is that of keeping people at work, thereby reducing the cost of sickness absence, which can be substantial.

Health surveillance of such groups generally takes the form of regular health examinations at predetermined intervals of, say, 6 or 12 months, according to the degree of risk involved. Such a system allows for early detection of evidence of occupational disease or condition and thus for its early treatment. Under the COSHH Regulations, health surveillance may be necessary where employees are exposed to, or are liable to be exposed to, a substance hazardous to health.

Under Reg. 5 of the MHSWR 'every employer shall ensure that his employees are provided with such health surveillance as is appropriate having regard to the risks to their health and safety which are identified by the [risk] assessment'. The ACOP accompanying the MHSWR indicates that health surveillance should be introduced where the assessment shows the following criteria to apply:

(a)  there is an identifiable disease or adverse health condition related to the work concerned

(b)  valid techniques are available to detect indications of the disease or condition

(c)  there is a reasonable likelihood that the disease or condition may occur under the particular conditions of work

(d)  surveillance is likely to further the protection of the health of the employees to be covered.

The primary benefit, and therefore the objective, of health surveillance should be to detect adverse health effects at an early stage, thereby enabling further harm to be prevented. In addition the results of health surveillance can provide a means of:

**5**

(a)  checking the effectiveness of control measures

(b)  providing feedback on the accuracy of risk assesssment

(c)  identifying and protecting individuals at increased risk.

The ACOP goes further by indicating that once it is decided that health surveillance is appropriate, such health surveillance should be maintained during the employee's employment unless the risk to which the worker is exposed and associated health effects are short term. The minimum requirement for health surveillance is the keeping of an individual health record. Where it is appropriate, health surveillance may also involve one or more procedures depending upon their suitability in the circumstances. Such procedures can include:

(a)  inspection of readily detectable conditions by a responsible person acting within the limits of their training and experience

(b)  enquiries about symptoms, inspection and examination by a qualified person such as an occupational health nurse

(c)  medical surveillance, which may include clinical examination and measurement of physiological or psychological effects by an appropriately qualified practitioner

(d)  biological effect monitoring, i.e. the measurement and assessment of early biological effects such as diminished lung function in exposed workers

(e)  biological monitoring, i.e. the measurement and assessment of workplace agents or their metabolites either in tissue, secreta, excreta, expired air or any combination of these in exposed workers.

The frequency of the use of such methods should be determined either on the basis of suitable general guidance (e.g. as regards skin inspection for dermal effects) or on the advice of a qualified practitioner. The employees concerned should be given an opportunity to comment on the proposed frequency of such health surveillance procedures and should have access to an appropriate qualified practitioner for advice on surveillance.

## OCCUPATIONAL HEALTH PRACTICE

The principal areas of occupational health practice, provided in the majority of cases by an internal or external occupational health service, can be classified as follows:

### Placing people in suitable work

It is of vital significance that workers be physically and mentally suited to the tasks they are required to undertake. Pre-employment medical examination, or health screening by an occupational health nurse, is a standard feature of the recruitment policies for many organisations.

Pre-employment screening activities should include not only an assessment of general fitness for the job but specific aspects of it, such as vision screening for drivers, VDU operators and people engaged in fine assembly work, the assessment of physical capacity and disability levels where heavy manual work is involved, and aptitude testing for a wide range of tasks. The starting point for pre-employment health screening normally takes the form of a health questionnaire (see Fig. 5.1 – Health questionnaire on p. 128) completed by the applicant which is assessed by the occupational health nurse as part of the pre-employment health examination.

### Health surveillance

Health surveillance concentrates on two main groups of workers:–

(a) those at risk of developing further ill-health or disability by virtue of their present state of health

(b) those actually or potentially at risk through the type of work they undertake during their employment.

### Providing a treatment service

The efficient and speedy treatment of injuries, poisonings and other ailments is an important feature of occupational health practice in that it prevents complications and aids rehabilitation. Feedback information from treatments following accidents and ill-health can also be used in improving prevention and control measures in the workplace.

## Primary and secondary monitoring

Primary monitoring is concerned largely with the clinical observation of sick persons who may seek treatment or advice on their condition. Such observation will identify new risks which were previously not considered. Secondary monitoring, on the other hand, is directed at controlling the hazards to health which have already been recognised. Typical forms of secondary monitoring are audiometry and various forms of biological monitoring, e.g. blood tests for lead workers.

## Avoiding potential risks

This is an important feature of occupational health practice with the principal emphasis on prevention, in preference to treatment for a known condition. Here the occupational health practitioner can make a significant contribution to the planning and design of work layouts, and to considering the ergonomic aspects of jobs and the potential for fatigue amongst workers. The effects of shift working, long hours of work and the physical and mental efforts of repetitive tasks would be taken into account in any assessment of the risks involved.

## Supervision of vulnerable groups

'Vulnerable' workers include young persons, the aged, the disabled, pregnant women and people who may have long periods of health-related absence. Special attention must be given to these persons in terms of counselling, assistance with rehabilitation and, possibly, in the reorganisation of their tasks to remove harmful factors. Routine health examinations to assess their continuing fitness for work is a standard feature of this form of health supervision.

## Monitoring for early evidence of non-occupational disease

Many occupational diseases are associated with specific occupations, e.g. silicosis with the pottery industry, byssinosis with the cotton industry. Whilst improvements in working conditions have greatly reduced the incidence of such diseases, routine monitoring of workers not exposed to these conditions is an important feature of occupational health practice. Here the principal objective is that of controlling diseases and conditions prevalent in industrial populations with a view to their eventual eradication. This form of monitoring also makes a great contribution to the control of stress-related diseases and conditions, such as mental illness and heart disease.

## Counselling

Counselling of an individual by an occupational health nurse, physician or health counsellor is a most important feature of occupational health practice. It may take two forms, namely counselling on health-related matters and counselling on personal, social and emotional problems. The availability of a sympathetic ear, independent of organisational controls, can assist the individual to come to terms with such problems more easily.

## Health education

Health education is primarily concerned with the education of staff towards healthier modes of living. It can also include the training of management and staff in their respective responsibilities for health and safety at work, in healthy working techniques and in the avoidance of health hazards.

## First aid and emergency services

This area includes the supervision of first aid facilities and ancillary equipment such as emergency showers, eye wash stations and emergency breathing apparatus, together with the preparation of contingency plans to cover major disasters such as fire, explosion or gassing accidents. This will entail the training of first aiders, in accordance with the requirements laid down in the Health and Safety (First Aid) Regulations 1981 and accompanying ACOP, rescue staff and key members of the management team in preparation for such disasters.

## Welfare amenity provisions

Specification and supervision of sanitation, hand washing and shower facilities, arrangements for storing and drying clothing, and the provision of drinking water, commonly feature in the routine monitoring aspects of occupational health practice.

## Environmental control and occupational hygiene

Control of the working environment and the environment outside the workplace are important features of occupational health practice. The employer has a duty under HSWA to provide a safe and healthy working environment for his employees by the recognition, measurement, evaluation and control of both short-term and long-term health hazards. He must also ensure that he does not expose people living in the vicinityof the workplace to health risks or public health nuisances from pollution of the air, ground, water, drainage system and water courses.

## *Liaison*

Occupational health practitioners liaise with a wide range of enforcement officers, such as medical and nursing advisers of the Employment Medical Advisory Service (EMAS), HSE inspectors, environmental health officers, planning officers and staff of the local Health Authority. They also liaise with local medical practitioners, keeping them informed of any health-related matters as far as they affect patients who may be employed by the particular organisation.

## *Health records*

The maintenance of appropriate health records relating to the health state of employees is an important feature of occupational health management and practice. The purpose of maintaining these records is to:

(a) assist occupational health staff to provide efficient health surveillance, emergency attention, health care and continuity of such care

(b) enable staff to undertake epidemiological studies to identify general health and safety problems and trends amongst employees and to identify problem areas and specific risks

(c) establish, maintain and keep up to date written information relating to people, hazards and current monitoring activities

(d) facilitate assessment of problems, decision-making, recommendations and the writing of reports.

The following records relating to individual employees, although not necessarily required by law, are recommended:

(a) initial and subsequent health questionnaire, interview, examination and screening test results

(b) relevant medical and occupational history, smoking habits, disabilities and handicaps

(c) attendance in the department for first aid, treatment, re-treatment, general health care and counselling

(d) injuries resulting from occupational and no-occupational accidents

(a) illness occurring at work or on the way to or from work

(f) sickness absences

(g) occupational conditions and diseases

(h) care and treatment provided

(j) advice given, recommendations and work limitations imposed

(k) referrals made to other specialists or agencies

(l) correspondence relating to the health of employees

(m) dispersal of cases following emergencies and treatment

(n) communications between occupational health staff and others, including written reports.

The following information should be included in occupational health records:

### (a) Personal identification details
Personal records are necessary for identifying and tracing individual employees and groups of employees exposed to particular risks. Identification details which are of particular value are:
   (i)  National Health Service number
   (ii)  National Insurance number
   (iii)  surname and forenames (maiden name, where applicable)
   (iv)  sex
   (v)  date, country and place of birth
   (vi)  usual address and date of taking up residence there.

### (b) Job history
Before he commences work with a new employer, an occupational history should be taken from the prospective employee. Details of the occupations in the current employment should appear on the individual record including transfers to alternative work with dates and duration in each job.

Other records which should be maintained include accident records, the results of work area visits, a daily attendance record of employees visiting the occupational health department or seen by the occupational health nurse and information relating to such matters as occupational health hazards, drugs, medical equipment and departmental procedures.

Regulation 11 of the COSHH Regulations requires that health surveillance be provided for persons who are, or are liable to be, exposed to a substance hazardous to health. Where such cases arise, the employer must ensure that a health record, containing particulars approved by the HSE, for each of his employees so exposed or liable to be exposed, be made and maintained. The record, or a copy thereof, must be kept in a suitable form for at least 30 years from the date of the last entry made in it. Where such an employer ceases to trade, he must forthwith notify the HSE in writing and offer these records to the HSE.

## OTHER AREAS OF OCCUPATIONAL HEALTH PRACTICE

### *Well women screening*

The Well Women Screening Service is available under the National Health Service. Two particular areas of importance are the early detection of cervical cancer and breast cancer. Both forms of cancer are known killers and represent a substantial number of deaths in the female population every year.

## Smoking, alcoholism and drug addiction

The relationship of cigarette smoking in particular to various forms of cancer is now well-recognised. Alcohol addiction can lead to chronic gastritis and possible inflammation of the intestines, and cirrhosis of the liver in some cases. Many organisations now have distinct policies covering both smoking at work and for dealing with cases where employees' level of performance and relative safety may be affected by alcohol abuse. Drug addiction implies that the user of the drug has developed a particular need or dependence on same in order to stay both physically and mentally normal. Once access to the drug is prevented or removed, certain physical and/or mental symptoms become apparent in the addict. In all the above cases, occupational health services have an important role in assisting individuals to 'kick the habit'.

## Hearing and eyesight deficiencies

It is a fact of life that, as people get older, so their ability to hear and see reduces. However, hazards can arise through people having imperfect hearing or vision and not recognising it. Audiometry and vision screen are now standard features of most occupational health service provided to organisations. The need for both audiometry and vision screening of certain groups of workers has been emphasised in the Noise at Work Regulations 1989 and the Health and Safety (Display Screen Equipment) Regulations 1992.

## MANAGING ABSENCE FROM WORK

Many organisations pay great attention to accidents at work and the cost, both direct and indirect, of these accidents. On the other hand, the management of absence is commonly neglected on the basis that the problem is insoluble and outside the control of management. However, absence from work on the part of staff and others can be a substantial cost to an organisation. It can take a number of forms:

(a) certificated or uncertificated sick leave, which may be of a long-term or short-term nature

(b) other forms of authorised absence, e.g. to attend training courses

(c) unauthorised absence and lateness.

Approximately 200 million working days are lost each year in the UK due to absence. This equates to 10–11 days absence per person per year, or a national absence rate of 4.6%. Many organisations may be totally unaware of their absence rate, whilst others may indicate rates in excess of the average 4.6%. Whatever the absence rate may be, sickness absence, in particular, represents a substantial continuing loss. It is essential, therefore, that the whole area of sickness absence be successfully managed.

## The legal position

Informal studies indicate that approximately 75% of the sickness absence in some organisations is taken by 25% of the work force. A substantial part of this absence is accounted for by people taking frequent periods of short-term sickness absence (1 or 2 days per month) which tends to go unnoticed. Long-term sickness absence, on the other hand, does tend to be noticed due, in many cases, to the need to recruit temporary labour or to require staff to work longer hours in order to cover that period of absence. However, it is only when attendance records are checked that the true scale of sickness absence is identified.

The legal issues are concerned with dismissal resulting from absence associated with:

(a)   failure to produce sick notes

(b)   frequent and short unconnected periods of absence

(c)   prolonged or continuous absence.

In the case of (a) and (b) above, it is essential for the organisation to go through a series of stages in order to counter a claim by an employee for unfair dismissal. In most cases an employee has a contract of employment, which would indicate a statement of terms and conditions of employment. He would be provided with details of any sick pay and the procedure for reporting sickness absence. An employee may be in breach of his contract of employment through either non-compliance with the scheme, e.g. failure to produce a sick note, or actual compliance with the scheme but with continuing frequent, short unconnected absences from work. In such cases, it would be appropriate for the company to withhold payment with a view to investigating whether there has been misconduct in these situations. (Procedures for dealing with such an investigation and its outcome are featured in many national agreements between employers and trade unions.)

Before an employee can be dismissed on the basis of absence, it must be established that the employer has behaved reasonably in dealing with the matter. For instance, he may have been interviewed by management or an occupational health nurse with a view to ascertaining the causes of his abscence, e.g. a sick wife or children. Where it can be established that there are no real grounds for his previously poor attendance record, and the matter has been brought to his attention both verbally and in writing, it would be reasonable for the employer to dismiss the employee.

Prolonged or continuous absence, on the other hand, hinges around 'capability' rather than 'misconduct'. Capability is assessed by reference to skill, aptitude, health or any other physical or mental quality. In many cases, perhaps through illness or injury, an employee may no longer have the capacity to undertake the work for which he is employed. For a fair dismissal decision to be reached, particularly where an employee complains of unfair dismissal to an Employment Appeals Tribunal, the employer should be able to answer satisfactorily the following questions:

1. Were offers of alternative employment considered?
2. Was the employee consulted sufficiently in terms of his problems, prospects and the possibility of dismissal?
3. Was the employee warned that further sickness could result in the possibility of dismissal?
4. Has there been a chance recently for the employee to comment on his health or capability to work?
5. Has the employer sought medical advice on the employee's condition?
6. Has the employer fully investigated all relevant matters relating to the dismissal decision?
7. How long, apart from illness, would the employment be likely to last?
8. Can the employer wait any longer for the employee to return? (Here a balance must be struck between the position of the employee, the interests of the organisation and the need to be fair.).
9. How important is it to replace the employee concerned?
10. Has the employee been consulted in the final step of the procedure?

Other questions which need consideration by an employer include:–

1. Were the terms of the employee's contract of employment, including sick pay provisions, fulfilled?
2. Did the nature of the employment constitute a key position?
3. What is the nature of the illness/condition?
   How long has it continued?
   What are the prospects of recovery?
4. What is the total period of employment?

If there is no adequate improvement in the attendance record, it is likely that, in most cases, the employer would be justified in treating the persistent absences as sufficient reason for dismissing the employee. At the completion of this deliberation stage, the employee must be advised of management's decision and the decision confirmed in writing.

Where an employee complains to an Employment Appeals Tribunal, an employer must provide the following information:

(a) evidence of a fair review of the employee's attendance record and the reasons for that record

(b) evidence of appropriate warnings being given, after the employee concerned had been given an opportunity to make representations.

## POLICIES ON SMOKING AT WORK

With the increasing attention that has been given to the risks associated with passive smoking, many organisations are considering the development of policies on smoking at work.

Smoking has been identified as a problem mainly in poorly ventilated open-plan offices and amongst employees who may suffer some form of respiratory complaint, e.g. asthma or bronchitis. Other people may complain of soreness of the eyes, headaches and stuffiness. In 1988 the HSE's booklet 'Passive Smoking at Work' drew attention to the concept of passive smoking, and work carried out by the Independent Scientific Committee on Smoking and Health identified a small but measurable increase in risk from lung cancer for passive smokers of between 10% and 30%.

The cost to industry of people smoking at work has never been measured in terms of time lost through smoking-related diseases and ill-health, fires caused through careless smoking and, in many cases, time wasted in indulging in the practice. On the other hand, for some people, smoking may be considered a minor form of addiction. Imposing a total ban on smoking, without consultation or assistance to give up the habit, could impose severe stress on these people. So how should companies seek to regulate this problem with a view to eventually phasing out smoking at work?

The first step is the development of a Policy on Smoking at Work. This should state the desire of the company to eliminate smoking in the workpace, the legal requirements on the employer to provide a healthy working environment, and that smoking is bad for the health of smokers and non-smokers alike. The organisation and arrangements for implementing the policy should request individual managers to support and implement the various stages of the operation, and the staff to comply with the policy.

On publication of the policy statement, the operation should proceed in clear stages, commencing with a questionnaire to all managers seeking information as to the number of employees on site, the number of smokers, possible problems anticipated through operation of an eventual ban, local efforts made to assist smokers to give up smoking, facilities available to smokers, and costs incurred in helping people to give up smoking, e.g. counselling, hypnotherapy.

The second stage is concerned with the provision of information, health education, therapy and counselling on the health risks associated with smoking with a view to raising the awareness of all concerned. Job applicants should also be advised of the policy at the interview stage. Some managers may be concerned at the risk of industrial action on the declaration of such a policy. However, in *Rogers* v *Wicks & Wilson* an employee, who subsequently resigned and claimed unfair dismissal when his employer announced a forthcoming ban on smoking, was held not to have been unfairly dismissed on the basis that employees do nto have a contractual right to smoke. Moreover, where such bans are introduced with sufficient warning and consulation with staff, an employer cannot be said to have acted unreasonably.

## CONCLUSION

The legal duty of employers to protect the health of employees and others who may be affected by their work activities is well established. However, in general, occupational health practices and the operation of occupational health services in the United Kingdom is at present restricted to a small number of organisations. While many managers would see the provision of occupational health services for employees as something of a luxury, the benefits to be derived can be substantial in terms of reduced operating costs, reduced losses through sickness absence and improved staff morale, resulting in better levels of performance and productivity.

---

### PRE-EMPLOYMENT HEALTH SCREENING QUESTIONNAIRE

Surname _____ Location _____

Forenames _____ Date of Birth _____

Address _____

_____

Tel. No. _____ Occupation _____

Position applied for _____

Names and address of doctor _____

_____

_____

_____

### SECTION A

Please tick if you are at present suffering from, or have suffered from:–

1.  Giddiness
    Fainting attacks
    Epilepsy

2.  Mental illness
    Anxiety or depression

3.  Recurring headaches

4.  Serious injury
    Serious operations

5.  Severe hay fever
    Asthma
    Recurring chest disease

6.  Recurring stomach trouble
    Recurring bowel trouble

7.  Recurring bladder trouble

8.  Stroke
    Heart trouble
    High blood pressure
    Varicose veins

9.  Diabetes

10. Skin trouble

11. Ear trouble or deafness

12. Eye trouble
    Defective vision (not
    corrected by glasses or
    contact lenses)
    Defective colour vision

13. Back trouble
    Muscle or joint trouble

14. Hernia/rupture

---

● **FIG 5.1  Health Questionnaire (see text p. 118)**

## SECTION B
Please tick if you have any disabilities that affect:–

| | | |
|---|---|---|
| Standing | Lifting | Working at heights |
| Walking | Use of your hands | Climbing ladders |
| Stair climbing | Driving a vehicle | Working on staging |

## SECTION C
How many working days have you lost during the last three years
due to illness or injury?                                   _____ days

Are you at present having any tablets, medicine or injections prescribed by
a doctor?                                         YES/NO

Are you a registered disabled person?                 YES/NO

## SECTION D

| Previous occupations | Duration | Name & address of employer |
|---|---|---|

_____

_____

_____

_____

## SECTION E
The answers to the above questions are accurate to the best of my
knowledge.

I acknowledge that failure to disclose information may require re-assessment
of my fitness and could lead to termination of employment.

Signature _____ Prospective employee      Date _____

Signature _____ Manager                   Date _____

# Loss Control Management – the TLC System

Accidents and occupational ill-health represent substantial losses to an organisation in both direct and indirect terms. Total Loss Control (TLC), a management system approach developed in the 1960s, is concerned with the prevention of these losses. It is defined as a programme designed to reduce or eliminate all accidents which downgrade the system, and which result in wastage of the organisation's assets. An organisation's assets are:

Manpower
Materials
Machinery
Manufactured goods
Money  (The five ('M's)

The evolution of the Total Loss Control programme may be shown thus:

**Injury Prevention**

+

**Damage Control**

=

**Total Accident Control**

+

**Business Interruption**
**(Fire, Security, Health and Hygiene, Pollution, Product Liability Losses)**

=

**TOTAL LOSS CONTROL**

Within the TLC concept some significant definitions may be arrived at:

*Incident –*
    An undesired event that could, or does, result in loss.
    *or*

An undesired event that could, or does, downgrade the efficiency of the business operation.

*Accident –*

An undesired event that results in physical harm or damage to property. It is usually the result of contact with a source of energy (i.e. kinetic, electrical, thermal, ionising, non-ionising radiation, etc.) above the threshold limit of the body or structure.

*Loss control –*

Any intentional management action directed at the prevention, reduction or elimination of the pure (non-speculative) risks of business.

*Loss control management –*

The application of professional management techniques and skills through those programme activities (directed at risk avoidance, loss prevention and loss reduction) specifically intended to minimise loss resulting from the pure (non-speculative) risks of business.

# LOSS CONTROL MANAGEMENT

## *Principal features*

### Planning

This includes forecasting, setting objectives, formulating policies, programming, budgeting and determining procedures.

### Organising

This aspect covers the structure of the organisation in terms of the allocation of responsibilities, the delegation of authority for certain actions and the relationships between people in terms of organising action.

### Leading

This is an important feature of the loss control management process. Leading implies clear decisions from senior management, the setting of the right example to subordinates on health and safety issues, motivating subordinates to pursue health and safety issues along with their main functions, selecting the right control strategies and the people to implement those strategies, and developing subordinates in the management of health and safety at work.

### Controlling

This is the fourth area of loss control management. It implies the setting of standards which are comprehensible to all concerned and achievable, measuring performance on a group and individual basis against the agreed standards, evaluating and, if necessary, correcting performance against agreed criteria.

Loss control management therefore involves the following:

(a) identification of risk exposure;

(b) measurement and analysis of these exposures;

(c) determination of those exposures that will respond to treatment by existing or available loss control techniques or activities;

(d) selection of appropriate loss control action based on effectiveness and economic feasibility; and

(e) management of programme implementation in the most effective manner subject to economic restraints.

A management control system implies:

(a) identification of work;

(b) setting of standards of work performance;

(c) the measurement of performance to specific standards;

(d) evaluation of performance in quantified form; and

(e) recognition of desired levels of performance and correction of substandard performance.

TLC recognises a number of fundamental truths and principles, including:

- *Principle of resistance to change* The greater the departure of planned changes from the accepted ways, the greater the potential resistance by the people involved.

- *Principle of definition* A logical decision can be made only if the real problem is first defined.

- *Principle of point of control* The greatest potential for control tends to exist at the point where the action takes place.

- *Principle of reciprocated interest* People tend to be motivated to achieve the results that you want to the extent that you show interest in the results *they* want to achieve.

- *Principle of reporting to the highest authority* The higher the level to which a loss control manager reports, the more management co-operation he achieves.

- *Principle of the critical few* In any given group of occurrences, a small number of causes will tend to give rise to the largest proportion of results.

- *Principle of recognition* Motivation to accomplish results tends to increase as people are given recognition for their contribution to those results.

- *Principle of future characteristics* An organisation's past performance tends to foreshadow its future characteristics.

- *Principle of multiple causes* A loss is seldom, if ever, the result of a single cause.
- *Principle of management results* A loss control manager tends to secure most effective results through and with others by planning, organising, leading and controlling.

## The role of the loss control manager

The role of the loss control manager, as a member of the management team, is to advise on measures which will reduce unit costs through the use of planned controls leading, ultimately, to minimised risk of accidents, incidents, fire and other incidents which downgrade the system.

On this basis, he must carry out two specific functions. First, he must locate and define areas involving incomplete decision-making, faulty judgement, administrative miscalculations, and clear evidence of bad management or individual work practices leading to incidents which downgrade the system. Second, he must provide advice to management as to how such mistakes can be avoided.

## STAGES OF TOTAL LOSS CONTROL

TLC is generally run as a programme over a period of, say, 5 years. The various stages are outlined below, and then looked at in more detail.

### Injury prevention

This stage is concerned with the humanitarian and, to some extent, legal aspects of employee safety and employers' compensation costs. It normally incorporates features such as machinery safety, cleaning and housekeeping procedures, health and safety training at all levels, personal protective equipment, joint consultation (safety representatives and safety committees), the promulgation of safety rules, regulations and disciplinary procedures, and safety propaganda.

### Damage control

This part of the programme covers the control of accidents which cause damage to property and plant, and which might conceivably cause injury. Essential elements of this stage are damage reporting, recording and costing.

### Total accident control

The stage of TLC is directed at the prevention of all accidents resulting in personal injury and/or property damage. Three steps towards total accident control are spot checking systems, reporting by control centres and health and safety audits.

## Business interruption

This entails the incorporation into the programme of controls over all situations and influences which downgrade the system and result in business interruption, e.g. fire prevention, security procedures, health and hygiene monitoring, pollution prevention, product liability procedures. Business interruption can result in lost money (e.g. operating expenses), lost time (e.g. cost of idle labour and equipment), reduced production and lost sales, perhaps through delays in delivery. Product liability claims are an increasing cost being borne by many manufacturers.

## Total loss control

This is the control of all insured and uninsured costs arising from any incidents which downgrade the system. It includes aspects associated with 'business interruption' and identifies the possible tools and methods of measurement.

The approach to health and safety at work varies substantially, from the typical injury prevention approach to the more financially-orientated approach used in TLC, as shown in Figure 6.1.

| Injury prevention | | TLC | |
|---|---|---|---|
| Humanitarian | 60% | Economic | 60% |
| Legal | 10% | Legal | 10% |
| Economic | 30% | Humanitarian | 30% |

● **FIG 6.1  Comparison between traditional injury prevention and TLC approaches**

## INJURY PREVENTION

This first stage of the programme incorporates many elements, of which the principal ones are outlined below.

### Machinery safety

Here the three basic rules apply:

(a)  if possible remove the hazard

(b)  if the hazard cannot be removed, guard it

(c)  if the hazard cannot be removed or guarded, ensure all personnel are warned of the hazard regularly.

(c) is not a perfect solution, and never will be, however. Emphasis must always be placed on (a) and (b).

### Cleaning and housekeeping

The philosophy here is that 'a clean plant is a safe plant'. An important management function is to maintain order and prevent disorder. In this case, 'order' is defined thus: 'A place is in order when there are no unnecessary things about, and when all necessary things are in their proper place.' The implementation of this part of the programme is assisted by regular cleaning and housekeeping inspections by individual managers.

### Rules and regulations

This aspect involves the use of rule books and/or health and safety manuals. It has been said that many company health and safety rules are 'written in blood', in that much safety improvement work is carried out, and health and safety rules written, as a direct result of accidents and occupational disease. On the other hand, there is a danger of health and safety rules becoming too unwieldy and onerous. Too many rules can be just as dangerous as too few and they must, therefore, be specific and appropriate to the people involved.

The following steps should be observed:

(a)  company health and safety rules should be presented in a form that is easily understood, e.g. a staff Health and Safety Handbook

(b)  rules should be logical, comprehensible and enforceable

(c)  rules should be known by all employees as a result of induction procedures, training programmes and discussion with foremen or supervisors

(d)  provision must be made and used for the enforcement of health and safety rules, i.e. with disciplinary procedures for 'unsafe behaviour'.

Rules and instructions must:

(a)  have significance, i.e. be related to an accident or injury situation, reduction in quality, etc.

(b)  be shown to have value to the person being instructed

(c)  provide a sense of security

(d)  have the acceptance of the work group.

### Health and safety committees

The basic function of a health and safety committee is to create and maintain an active interest in health and safety with a view to reducing occupational ill-health and accidents. This is achieved by:

(a)  discussing and formulating specific policies on health and safety issues and recommending their adoption by management

(b)  discovering unsafe conditions and practices and determining remedial action

(c)  working to obtain results by having its recommendations approved by management and put into practice

(d)  teaching basic aspects of health and safety practice to committee members who will, in turn, teach them to all personnel in the organisation.

One of the problems of health and safety committees is that they are frequently seen as a management-run 'talk shop' which never achieves anything. To overcome this view:

(a)  the role and function of the chairman, secretary and members should be clearly identified in the Statement of Health and Safety Policy

(b)  the constitution should be formally established in writing

(c)  an agenda and minutes should be produced for each meeting and publicised throughout the organisation

(d)  there should be clear-cut evidence of committee decisions being implemented within the time scale laid down at the meeting.

### Contests, competitions, award schemes, etc.

These activities are directed at raising the profile of health and safety within the organisation. As such they need very careful monitoring and control if they are to be successful in changing the attitudes of people towards the adoption of safe working practices. The actual extent of the various contests, competitions and award schemes is varied. It can include competitions based on housekeeping standards or the wearing of personal protective equipment, safety poster and slogan contests, raffles, etc. Under no circumstances, however, should competitions or award schemes be based on accident incidence rates as, inevitably, this will result in a reduction in accidents reported in order to win the award. The best form of contest is one based on a form of safety monitoring, such as inspections or safety sampling exercises.

### Personal protective equipment

As with machinery safety, if more consideration were given to the design of work equipment and processes, there would be less need for personal protective equipment. A good example of this is noise pollution. As better techniques for noise reduction or elimination are developed through system engineering and human factors engineering, the need for hearing protection, e.g. ear muffs, ear plugs, will also be reduced or eliminated.

## Deficiencies in the injury prevention approach

Studies based on the TLC concept indicate that the injury prevention approach has a number of limitations and deficiencies. These include:

### Two narrow a base of action

Progress is often hampered by tunnel vision on the part of management and safety specialists, as shown by the pre-occupation with preventing injury, but not all accidents, though these also downgrade the system. For instance, there is a need to examine aspects of total environmental control, e.g. in-plant safety, off-the-job safety, fire protection, air and water pollution control, environmental hygiene and product safety.

### Etiological causes bypassed
In some case the underlying causes of accidents are overlooked. As a result, corrective actions have been directed at symptoms rather than causes.

### Patchy prevention activities
Corrective efforts frequently directed at incidents involve a series of 'fire-fighting' operations, whereas what is needed is in-depth direct action to prevent recurrence. Furthermore, there has been too much emphasis on post-accident strategies, such as accident investigation, and too little on pre-accident strategies, such as control of the working environment.

### Universal antidotes
The 'broad brush' approach has been, and is, a waste of time and effort, listing approaches that will have no beneficial results in the environment in which they are used, or putting more effort into a situation than is required. Ninety per cent of all injuries occur as a result of unsafe work habits, wrong attitudes and personal risk taking, yet 90% of the safety specialist's time, in many cases, is directed at unsafe 'things', such as poor machinery guarding, uneven floors, etc., with incorrect operator attitudes being totally neglected. This general inattention to the human factors side of occupational safety is now better recognised by management and enforcement agencies alike.

### Limited formal training of safety practitioners
There is no legal requirement for a safety practitioner to be trained and far too many are practising in the dark due to insufficient formal training. Furthermore, many safety practitioners have been appointed in default, and many operate at too low a level within the organisation to be effective, not being accepted as a member of the management team.

### Little applied research
Any research into injury prevention has been too little and very thinly spread.

## DAMAGE CONTROL

Accidents commonly result in property damage as well as human injury. TLC examines not only injury accidents but also those resulting in damage to property, plant and equipment. In the TLC context an accident is defined as 'an unintentional or unplanned happening that may or may not result in property damage, personal injury, work process stoppage or interference, or any combination of these conditions, under such circumstances where personal injury might have resulted'.

This definition provides an excellent guide for determining whether or not the damage involved was 'accidental', and whether it should be included in the all-accident control programme. Two questions must be asked:

1. Was the happening 'unintentional' or 'unplanned?'

2. Was there a realistic possibility of personal injury being involved?

If the answer to each of these questions is 'Yes', the cost should be classified as accidental property damage.

## Putting damage control into operation

Once it is decided that incidents involving property damage are to be included in the programme, through the establishment of damage control centres, the organisation will have, in effect, moved towards the Total Accident Control stage.

Three basic steps are essential to introduce Damage Control into the safety programme, once the control centres have been established.

### (a) Spot checking
This involves the spot checking of repair centres, making observations and taking notes. It permits estimates of damage to be verified by comparing the costs of repair by sample observation. Great emphasis should be placed on the potential for personal injury resulting from property damage accidents.

### (b) Reporting by control centres
This is a system whereby the repair control centre and key personnel report property damage situations. The system should be simple, designed to operate with a minimum amount of written reporting, and flexible since repair cost accounting methods may vary from workplace to workplace.

### (c) Auditing
During the first two steps towards completing integrated reporting, some modified form of auditing should be maintained. The third and final step to a completely effective reporting programme is more complete auditing. One way of achieving this objective is for the health and safety specialist to request copies of work orders processed through the maintenance planning organisation and cost control centre. On-the-spot investigations, to determine whether or not accidental damage was involved in these orders, are then undertaken by the health and safety specialist.

The investigation of property damage accidents serves a number of purposes, namely:

(a)   it makes supervisors more aware of the cost involved in all accidents and of their responsibilities in controlling them

(b)   it eliminates the common causes of personal injury and associated accidents

(c)   it eliminates judgement determination and accelerates the investigation procedure

(d)   it provides a source of valuable data for detailed analysis

(e)   it assures management of specific remedial action to prevent recurrence of damage.

## BUSINESS INTERRUPTION

'Business interruption' aspects of a TLC fundamentally include those insured and uninsured costs arising from any incidents which downgrade the system. These costs include those associated with injuries, property damage, fire, security, occupational health and hygiene incidents, pollution and product liability, which are the main areas of loss within the TLC concept.

Business interruption results when any work activity is carried out in such a way that the ability of the organisation to provide goods and / or services is adversely affected. For practical purposes, each organisation must define the parameters of what constitutes 'business interruption' within that organisation. Industrial injury, occupational disease, fire or explosion, pollution of air, ground or water, product losses, equipment and vehicle damage, security violations and product liability incidents are all cited as high cost examples. They cause the major interference with expected productivity and commonly result from errors or mistakes in judgement or action taken. Their prevention or avoidance is a primary management responsibility. They are best avoided in a management climate which encourages a pride in performance and does not knowingly allow errors to occur. Business interruption costs can be split into four main areas:

**Money commitment**
Money losses can include:

(a)   losses of capital through bad planning, incorrect decisions and replacements

(b)   operating expenses in excess of normal

(c)   litigation as a result of negligence.

**Time commitment**
Time costs can include:

(a)   idle labour, often as a result of accidents resulting in injury or damage to plant, machinery and facilities

(b) idle plant, due again to injury and property damage accidents.

### Reduced production

Losses as a result of reduced production can be caused by:

(a) the need for alternative courses of action, which may be affected by a lack of decision-making by management

(b) the need for alternative sources of supply, which can be subject to problems with the availability of items and ordering arrangements

(c) the need for alternative transportation caused by vehicle breakdowns and losses

(d) the need to re-run or re-work production operations as a result of poor quality inspection and control, and inadequate supervision during manufacture.

### Sales lost

There are many reasons for the loss of sales. Two principal causes are:

(a) delays caused by failure to meet delivery dates, resulting in loss of good faith

(b) alienation created by a poor public image associated with a high level of accidents.

## Elements of business interruption

The principal elements of this stage of the TLC programme are fire prevention and control, security, occupational health and hygiene, pollution prevention and control and product liability.

### Fire prevention and control

Accident prevention and fire prevention are so closely allied that, in loss control evaluation, it is standard practice to incorporate fire prevention and control as an essential party of the programme. Fire prevention and control programming can be divided into Fire Prevention and Fire Extinguishment.

### *1. Fire prevention*

This incorporates a number of features, including the design of fire detection equipment, the maintenance of all facilities, as well as fire appliances, with the principal objective of preventing a potential major fire from starting. Fire prevention measures include segregation of areas, specific procedures covering potential fire hazards in processes such as welding, and inspection of equipment, facilities and systems on a frequent basis.

## 2. Fire extinguishment

The organisation of fire control groups, training of staff in extinguishment techniques and the appointment of fire officers to oversee these activities is an important feature of a fire control strategy. Fire control also entails the provision of well-maintained, regularly-serviced and correctly-located fire appliances and the identification and marking of high fire-risk areas.

## Security

The actual losses to organisations from theft, pilfering, vandalism, industrial and commercial espionage, bomb and bomb scares is unknown. Broadly, security procedures should ensure adequate protection of buildings, personnel, equipment, funds and confidential data against sabotage, theft, pilfering, vandalism and other activities which might endanger or interrupt normal operations. To achieve a high level of business security requires a well-developed plan, sound procedures, adequate physical controls, trained and physically fit security staff and modern equipment. A well-developed security programme should incorporate the following elements:

## 1. Security plan

Before any form of security plan is prepared, the actual need for such a plan should be analysed. This can only be done through a critical survey of the current situation to assess areas of potential loss, possible threats to the success of the business and general vulnerability, e.g. as a result of theft, loss of confidential data, etc. One of the main factors in developing a security plan will be the cost of administering, organising, supervising and training staff for the plan.

## 2. Procedures

Once the plan has been formulated, it becomes necessary to establish procedures in a clear and concise manner. Such procedures should include a general policy on security, an organisation chart indicating levels of responsibility for security, employment practices for the hiring of security staff, including detailed job descriptions and the duties and responsibilities of such staff, such as arrest and search powers. Detailed systems should also be developed to ensure the security of premises, protection of property, control of personnel and vehicles, including traffic control, security liaison with the fire control function, accident control and first aid to the injured.

## 3. Physical controls

Control over entry is a prime requisite in any security programme. Barriers at perimeters should be of sound construction and well maintained. A high level of external and perimeter lighting should be maintained, particularly in critical areas, and critical equipment, such as plant valves, regulators, transformers, etc. should be protected against tampering. Physical openings in the fabric of buildings, such as windows, air ducts, tunnel entrances and sewers, should be protected against entry.

### 4. Personnel

There is a case for some form of security screening of staff, together with a practical system for the identification, admission and control of employees and visitors. Effective control over vehicles should include the inspection of cars, trucks and other vehicles entering of leaving the premises, together with any packages taken into or out of the premises in such vehicles, and the regular inspection of vehicle parking areas. Control may include the issuing and display of vehicle passes. Classified information must be carefully controlled. In this case, such material should be stored in lockable filing cabinets, and access to vital records should be limited to as few people as possible. The guard force should be adequate to be able to maintain this level of protection.

### 5. Equipment

Depending upon the type of premises, the goods manufactured and the vulnerability to espionage, theft, sabotage, etc., consideration should be given to the installation of intrusion detection devices, closed circuit television, adequate gates and gatehouses, two-way radio communication, special multiple lock controls, emergency lighting systems and other automatic devices to deter potential intruders.

Disaster and emergency control programmes should be included in the security plan. Disaster or emergency control refers to all measures taken by an organisation to:

(a)  assure the uninterrupted productive capability of a facility

(b)  minimise the loss or disruption of production from any hazard

(c)  rapidly restore production following a disaster.

### Occupational health and hygiene

Occupational health and hygiene practice covers a very broad area. It is concerned with the protection of people's health at work (occupational health) and the identification, measurement, evaluation and control of contaminants and other physical phenomena, such as noise and radiation, which could otherwise adversely affect the health of people at work (occupational hygiene). This field covers many areas, such as the control of radioactive substances, lasers, hazardous substances and various forms of airborne contamination (dusts, fumes, gases, vapours, mists). It is also concerned with various forms of health surveillance, such as pre-employment health screening of staff, regular health examinations, counselling on health-related issues and, in certain cases, biological monitoring (blood and urine tests, for instance).

The occupational health and hygiene components of a TLC programme should incorporate the following elements:

### 1. Identification/recognition

Identification of health-related problems amongst staff, which may have been created by the working environment, requires in many cases a team approach by specialists, such as occupational physicians, health nurses and hygienists, supported by health and safety practitioners, engineers and trade union safety representatives. These specialists should be aware of the various stressors – physical, chemical, biological and psychological – which could affect people at work.

### 2. Measurement
Once the stressor has been identified, its magnitude should be measured. In the case of airborne contaminants, this may entail personal dose-monitoring or some form of static sampling to determine the concentration in air of the contaminant. Measurement should fundamentally identify the intensity of exposure and duration of exposure to the identified stressor.

### 3. Evaluation
Following measurement of the individual stressor, evaluation can be made against known hygiene standards, e.g. Occupational Exposure Limits, noise action levels specified in the Noise at Work Regulations 1989.

### 4. Prevention or control
In the case of substances identified as hazardous to health, the Control of Substances Hazardous to Health Regulations 1988 (COSHH) require that exposure to them must be either prevented or controlled. Prevention of exposure can be achieved by either a prohibition on the use of the substance, total enclosure of the source or by substituting a less hazardous substance. Control, on the other hand, may entail an engineering solution, such as the installation and operation of local exhaust ventilation systems, or by changing the process to afford better operator protection. In relatively low risk situations the provision and use of specific items of personal protective equipment may afford the appropriate level of operator protection. In all cases, prevention and control strategies should be supported by the provision of information, instruction and training for staff and adequate supervision.

## Pollution
A pollutant is defined by the World Health Organisation as anything which may affect Man's physical and mental well-being. This implies the establishment of tolerance levels for a wide variety of physical, chemical and biological contaminants which, if exceeded, result in pollution. Within this context of 'pollution' can be included pollution of the air, ground and water supplies, and other forms of pollution, such as noise.

### 1. Air pollution
Air can become polluted through the emission from industrial and domestic fuel-burning appliances of gases, such as sulphur dioxide and smoke. There may also be gas emissions from vehicles. Other forms of air pollution

include, for instance, the emission of grit and dust from cement manufacturing processes and of oil smuts from oil-fired industrial processes.

## 2. Water pollution
A wide variety of water pollutants can be encountered, such as sewage, detergents, chemical effluents, etc.

## 3. Ground pollution
Ground can become polluted through the depositing and disposal of toxic wastes, pesticides, industrial and domestic refuse, asbestos waste, etc.

## 4. Noise pollution
'Noise' is simply defined as 'unwanted sound'. Exposure to noise can result in occupational deafness (noise-induced hearing loss), fatigue, speech interference and stress. In some cases, noise can be a contributory factor in industrial accidents. Noise nuisance from, for instance, industrial processes, can result in fatigue, stress and a range of psychosomatic symptoms.

Prevention or control of the above forms of pollution is an important feature of the TLC programme. Apart from indicating generally a waste of resources and poor standards of preventive maintenance, pollution incidents or continuing pollution situations, which may affect local residents and/or the surrounding communities, are bad for the organisation's public image and can have an indirect effect on purchasing decisions made by potential customers.

## Product liability
Product liability refers to a system whereby a producer (manufacturer, designer, importer, etc.) may incur both criminal and civil liability for a defective product. Criminal law expects the producer to do all that is reasonably practicable to ensure that his products are safe when used properly. It is not necessary for injury to occur. Proof of risk of such injury is sufficient. Civil liability varies according to whether the product causes injury or not. Typical situations with a potential to cause loss and/or injury to purchasers and users of products are outlined below.

## 1. Impact
A product may come apart or disintegrate resulting in death or injury to persons in the immediate vicinity, e.g. an abrasive wheel bursting.

## 2. Fire and/or explosion
The advent of new materials and the development of new uses for older materials gives this particular risk more importance. Furthermore, many liquids are packaged under pressure with liquid gases in a variety of containers which can rupture, e.g. aerosols.

### 3. Radiation

Some man-made isotopes and naturally-occurring materials emit ionising radiation, all of which can be harmful in excessive doses.

### 4. Compatibility

Certain substances may be incompatible in the presence of, or when mixed with, other substances, resulting in explosion, fire and health-related risks.

### 5. Relative toxicity

Certain substances may be classified as toxic, corrosive, harmful or irritant and, depending on the nature of the substance, can result in poisoning or produce harmful effects, including death.

### 6. Defective design and assembly

At the design stage of products all possible failure and misuse situations should be considered. Incorrect or defective assemblies can result in death, injury and various forms of loss.

### 7. Inadequate instructions

A lack of warnings to users as to the circumstances of use, and limitations in use, or of specific directions to ensure correct assembly, has proved to have disastrous consequences with many products.

### 8. Applications other than intended

The easier the visualisation of additional applications of a product, the greater is the burden on the manufacturer to inform users of the risks of incorrect use or use for a purpose other than specified.

### 9. Absence of available safeguards

By law certain guards and safety devices must be fitted to a variety of work and other equipment. Where such guards and devices are not provided as with, for instance, an imported machine, the importer/supplier could be liable under criminal law, e.g. the Health and Safety at Work Act 1974.

## LOSS CONTROL MANAGEMENT

The principal elements of loss control management, as compared with other forms of health and safety management, are summarised below.

### Identification and classification

Loss control management can be directed at an inter-related group of problem areas that could include, but not be limited to, occupational health and safety (personal injury, ill-health and property damage), environmental health, security, fire loss control and product safety.

Management activities required to produce the desired control of loss in any of these areas could include:

(a) monitoring and investigation procedures

(b) emergency procedure

(c) rules, regulations and working practices

(d) physical protection

(e) supervisor, staff and professional training

(f) skill training

(g) general promotion activities

(h) job observation

(i) reparative action

(j) salvage operation

(k) design and maintenance engineering

(l) purchasing standards and practices

## *Loss measurement*

Measurement tools for loss control managers can be placed in three categories – consequence, cause and control.

### (a) Consequence
This could include major, serious, recordable and minor losses, or classifications of personal injury, property damage and other losses in terms of frequency and severity rates. Measurement of consequence could also be expressed as actual or potential loss rates.

### (b) Cause
This could take the form of actual or potential cause rates. The typical loss analysis of accident causes related to acts, conditions or management deficiencies that resulted in loss would best represent the actual type of cause measurement. This measurement category could be related to other types of loss as well. The result of sampling procedures applied to physical conditions or safe employee behaviour could be classified as potential cause measurement.

### (c) Control
The first step in utilising this important measurement class is to clearly define management's general activities as described earlier. Safety work would vary from organisation to organisation depending upon the degree of programme sophistication.

Assuming that a local standards or policy has been established for management's input into each of these activities, the remaining key step is to establish a method to quantify the degree of management effort in each area. This may be by specific sampling techniques, random sampling or interviews. A key to homogeneity and accuracy is the degree of objectivity designed into the specific method of measurement used.

## THE COST-BENEFIT APPROACH TO HEALTH AND SAFETY

The cost-benefit approach, as opposed to a legalistic one, i.e. of ensuring compliance with the minimum standards imposed by the law, is a significant feature of loss control management.

### *The costs*

All accidents, incidents (which do not necessarily result in injury), and cases of occupational ill-health represent some form of financial loss to the organisation. These costs to the organisation can be classified as direct and indirect costs.

### Direct costs

These include:

(a)  claims in the civil courts made by injured employees and others, e.g. contractors, members of the public, or those who may have contracted occupational ill-health, as a result of exposure to hazards in the workplace

(b)  insured costs

(c)  employer's liability costs

(d)  fines imposed by the criminal courts for breaches of health and safety legislation.

### Indirect costs

Whilst the direct costs of accidents, incidents and occupational ill-health can be readily identified, the indirect costs are commonly overlooked and get lost in the operating costs of the business. The indirect costs can be classified as follows:

(a)  medical, first aid and transport to hospital

(b)  lost time of the injured employee, plus loss of that individual's skills

(c)  lost time of other persons – management, supervisors, employees

(d)  replacement labour, including training and retraining costs

(e)  welfare payments to injured employees

(f)  loss of production – injury and property damage incidents

(g)   repair or replacement of damaged plant, equipment, raw materials and finished products

(h)   increased health and safety administration costs

(i)   other costs – court preparation, administration, legal fees, expert witness fees, etc.

(j)   loss of employee morale leading to reduced productivity

(k)   'critical incident' costs, i.e. where there is no injury or damage to plant, but which result in process and plant stoppage

(l)   activities of trade union safety representatives following accidents, incidents, occupational ill-health or scheduled dangerous occurrences, e.g. refusal to operate certain items of plant or use certain substances until clearance given by independent expert or HSE inspector, extra time spent in discussion and consultation.

## *The benefits*

The benefits to be derived from implementing a TLC approach to health and safety management are listed below:

Improved industrial relations

Improved morale and operator performance

Reduced labour turnover

Reduced absenteeism

Reduced error rates

Reduced training costs

Reduced plant maintenance costs

The reduced accident level and improvement in operator performance should result in increased productivity and profitability for the organisation.

## CONCLUSION

TLC, as a series of integrated health and safety management systems and approaches, has much to offer organisations. The principal objective of TLC is to achieve management commitment to safety, health and welfare, the main emphasis being on the fact that all accidents and incidents which downgrade the system represent losses to the organisation. By preventing or controlling these losses, whether resulting in injury, property damage or business interruption, considerable cost savings can be achieved.

Total Loss Control is a management-orientated system. It aims at identifying responsibility for control of health and safety, motivation of managers towards better health and safety performance and the elimination of losses associated with poor standards of this performance.

# Risk Management

Over the last thirty years there has been a trend away from the traditional injury prevention approach to health and safety at work to one concerned with preventing and controlling all forms of loss to an organisation as with the techniques of Total Loss Control. Much of the drive for bringing about this change has come from the insurance world.

## WHAT IS RISK MANAGEMENT?

Risk management, as a management science, could be said to take the TLC concept a stage further. It may be defined in a number of ways:

1. The minimisation of the adverse effects of pure and speculative risks within a business.
2. The identification, measurement and economic control of the risks that threaten the assets and earnings of a business or other enterprise.
3. The identification and evaluation of risk and the determination of the best financial solution for coping with the major and minor threats to a company's earnings and performance.
4. A technique for coping with the effects of change.

Broadly, risk management techniques aim at producing savings in insurance premiums by first defining and then minimising areas of industrial and other risk. It seeks not to discredit insurance arrangements but to promote the concept of insuring only what is necessary in terms of risk. On this basis, the manageable risks are identified, measured and either eliminated or controlled, and the financing of the remaining or residual risks, normally by insurance, takes place at a later stage.

Generally, organisations do not assess their real needs for insurance cover and measure these needs against their own ability to take on the risks. Instead, they treat insurance more like a tax which must be paid and never actually assess what they are buying. On this basis, the organisation buys insurance against probabilities or likelihoods based, perhaps, on their previous claims history, whereas it should really be covering unlikely or unforeseeable risks which it cannot control. Furthermore, there is a tendency to examine only those risks which insurance companies traditionally cover, such as fire, and to ignore other risks which could be significant financially should they arise – the 'tunnel approach'. What risk management requires from company management is time to consider and evaluate the risks, discipline and systems for controlling the identified risks they are capable of controlling, and the acceptance of a certain amount of covered risk. In other words, managements need to accept the fact that the handling of risks is much wider than just the purchase of insurance. They need to examine all the risks, evaluate them and treat insurance as a secondary consideration rather than the starting point.

## Categories of risk

Risks can be split into two main categories, namely catastrophic risk, which demands insurance, and risks associated wth wastage of the organisation's assets. The latter is where the scope of self-insurance and diminution of risk is most evident, and is why organisations appoint risk managers and sometimes establish risk management subsidiaries.

Risks may also be pure risks or speculative risks. Pure risks can only result in a loss to the organisation. Speculative risks may result in either gain or loss. Within the context of a risk management programme, risk may be defined as the chance of loss, and the programme is therefore geared to the safeguarding of the organisation's assets, viz. manpower, materials, machinery, methods, manufactured goods and money.

## THE ROLE OF RISK MANANGEMENT

The role of risk management in commerce and industry is to:

(a) consider the impact of certain risky events on the performance of the organisation

(b) devise alterantive strategies for controlling these risks and/or their impact on the organisation

(c) relate these alternative strategies to the general decision-making framework used by the organisation.

The process commences with the identification and analysis of any particular risk. The risk is then assessed or measured in terms of cost to the organisation and the measures to be taken to eliminate or control it are then implemented. For most organisations the problem of risk will not only relate to temporary business interruption, but also to future earnings and cash flow, together with damage to fixed assets. Fire is, perhaps, the most commonly-encountered risk. In this case it would be necessary to measure what the loss to any particular asset would be in terms of lost production, finished products and raw materials, and any consequential effects, such as the company's potential for regaining its market share once rebuilding and reinstatement of manufacturing capacity is finished.

Other areas that can affect earnings and cash flow include marketing, staff, technical and even political risks. In the case of marketing risks, there is always the risk of sales resistance to a particular product, perhaps as a result of adverse publicity following a product liability incident, or the risk of obsolescence due to changing tastes, service needs or the introduction of a better but comparable product by a competitor. Technical risks could involve missed performance targets due to plant and equipment breakdown, failure in a given production run or delays in receiving components manufactured outside the organisation. Labour and political risks may arise from time to time. Recruiting the right sort of labour force, in terms of skill, knowledge, experience and commitment, is one risk. The risk of government intervention, e.g. nationalisation or specific legislation relating to the product or service, is a further risk, together with official government financial policies which may seriously affect dividends and earnings. Certain health and safety-related risks could also be included, such as increased public awareness following a major incident, like the Flixborough incident. This could result in closure of the manufacturing base, should the risks to members of the public be too great.

Having identified the risks to the organisation and its assets, the function of risk management is to determine the probability or likelihood of the risks arising and the severity of the outcome in each case. It could follow that the risk could be controlled, in the case of a fire risk, by the installation of a better fire protection system, e.g. sprinkler system, automatic fire alarm, etc. and improved safety precautions. The implementation of these measures frequently results in some form of savings in the long term, such as insurance premium reductions, cash grants and tax allowances.

Where it is not possible to exercise economic control over an identified risk, then the organisation must either take the loss as an operating cost or insure against the risk. Thus the financial outcome or consequences of risks must be considered and a decision made as to how to meet these financial liabilities if and when they occur. The usual procedure is to cover these liabilities through the insurance market, which may not necessarily be cost-effective in the long term.

## The risk management philosophy

The trend diagram overleaf (Fig. 7.1) can be used to illustrate the basis of the risk management philosophy or concept. The solid line in the diagram depicts the investment in safety, whereas the broken line indicates the costs of accidents. As such, the diagram shows the historic trend of safety and accidents, in which it can be seen that as the investment in safety measures increases, so the accident rate decreases. The diagram indicates a very important point, namely that the risk cannot be reduced to zero. The safety investment curve is really an asymtotic curve; this will make it impossible for the accident risk curve to reach zero. The shaded areas in the diagram are meant to represent the risk areas in which people work, the left-hand shaded area being the case in the past, compared with the right-hand shaded area, which would represent the current situation. It will be noted that as this area moves across the diagram from left to right, so the width of the area gets smaller. It is interesting that legislation recognises that there is no absolute way in which to define safety, and terms such as 'so far as is reasonably practicable' and 'suitable and sufficient measures' commonly qualify legal duties. Legislation, as it becomes more refined, effectively reduces the room for manoeuvre on the part of organisations.

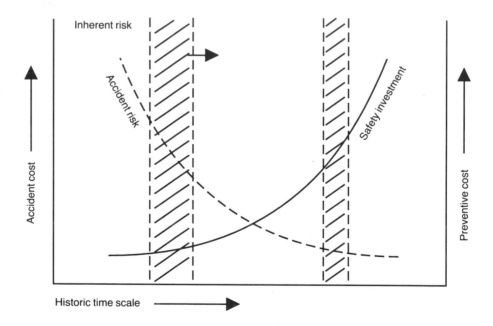

● **FIG 7.1  Risk management trend diagram**

Where an organisation is conscious of its health and safety responsibilities and has a real will to discharge these responsibilities to the best of its ability, it would tend to work towards the left of the area. On the other hand, a company that is doing no more than merely ensuring compliance with the law, i.e. the minimum effort possible, will tend to work towards the right of the area. Whether such an organisation knows it or not, it is in fact accepting a higher accident risk. One of these accidents may cause a loss which far outweighs the lesser expense on safety measures taken.

# THE RISK MANAGEMENT PROCESS

As stated earlier in this chapter, risk management involves the identification, evaluation and economic control of risks within an organisation. The risk management process, therefore, incorporates the following elements.

### Identification of the exposure

This entails a survey of all areas within the operation which could be sources of fortuitous risk. It requires a knowledge of all the assets, both tangible and intangible, the sources of earnings, both direct and indirect, plus an awareness of the potential liability which can be imposed by both common low and statute law. Since assets, earnings and legal requirements are all subject to constant change, the search for exposure to risk is a constant process of evaluation and prediction.

The identification of risk may be achieved by a multiplicity of techniques, including physical inspections, management and worker discussions, safety audits, job safety analysis and Hazard and Operability (HAZOPS) studies. It can also involve the study of past accidents to identify areas of high risk. Typical risks include those associated with:

(a) fire, flood, storm, impact, explosion, subsidence and other hazards

(b) accidents and the use of faulty products

(c) human error, e.g. loss through damage or malfunction caused by mistaken operation of equipment or wrong operation of an industrial manufacturing programme

(d) theft and fraud

(e) contravention of social or environmental legislation

(f) political risks (the appropriation of foreign assets by local governments, or of barriers to the repatriation of overseas profits)

(g) computer fraud, viruses and espionage

(h) product tampering situations and malicious damage.

### Analysis and evaluation of the risk

Risk analysis and evaluation (or measurement) may be based on economic, social or legal considerations. What is the likelihood of the risk arising? How frequently will the loss occur? What is the expected severity of the loss?

Economic considerations should include the financial impact on the organisation of the uninsured costs of accidents, the effect on insurance premiums, the overall effect on the profitability of the organisation and the possible loss in production following, for instance, enforcement action by the authorities.

Social and humanitarian considerations should include the general wellbeing of employees, the interaction with the general public who either live near the organisation's premises or come into contact with the organisation's operations, e.g. transport, noise nuisance, effluents, and the consumers of the organisation's products or services. Legal considerations should include possible constraints from compliance with health and safety legislation and other legislation concerning fire prevention, pollution and product liability.

The probability and frequency of each occurrence, and the severity of the outcome, including an estimation of the maximum potential loss, will also need to be incorporated in any meaningful evaluation of risk.

### Risk control

Once the risk has been assessed and evaluated, managers must then decide on appropriate action. Risk control is a feature of many managers' responsibilities. The production manager must avoid shutdowns of the production process, the company lawyer must protect the organisation against, for instance, libel, the business manager must protect the assets from loss and theft. There is a need, therefore, for full co-ordination of these efforts with a view to assuring management that all exposures have been identified and the financing of the risk is at the lowest cost.

Whilst feedback from previous loss situations provides information, the control of risk should be approached in anticipation of future problems. Risks are inherent in the safety of the work force, the maintenance of property, protection of the environment, the security of people and the product. A comprehensive risk management programme that is fully and correctly co-ordinated will make it possible to eliminate or reduce many risks.

### Financing of the risk

The loss potential associated with some risks is too small to affect the financial stability of the organisation. Such risks can be safely ignored. There will be other risks whose loss frequencies are statistically stable and which can be considered a regular cost of being in business. Typical risks in this category are those associated with the operation of a fleet of vehicles, which can be predicted within an acceptable range, and which can be budgeted for as part of the cost of being in business. The final risk category incorporates those risks, i.e. catastrophic risks, that are too large, or that occur too infrequently, to be ignored or treated as an operating expense.

## Risk control strategies

### Risk avoidance

This strategy involves a conscious decision on the part of the organisation to avoid completely a particular risk by discontinuing the operation that produces the risk. It presupposes that the risk has been identified and evaluated. A typical risk avoidance strategy would be the decision to cease to use any form of asbestos, replacing it with a safer alternative.

### Risk retention

The risk is retained within the organisation where any consequent loss is financed by the company. There are two aspects to consider here – risk retention with knowledge and risk retention without knowledge.

(a)   Risk retention with knowledge

This covers the case where a conscious decision is made to meet any resulting loss from within the organisation's financial resources. Decisions on which risk to retain can only be made once all the risks have been identified and effectively evaluated.

(b)   Risk retention without knowledge

This usually results from lack of knowledge of the existence of a risk or an omission to insure against it. It often arises because the risks have not been either identified or fully evaluated.

### Risk transfer

Risk transfer refers to the legal assignments of the costs of certain potential losses from one party to another. The most common way of effecting such transfer is by insurance. Under an insurance policy, the insurer undertakes to compensate the insured against losses resulting from the occurrence of an event specified in the policy, e.g. fire, accident.

### Risk reduction

The principles of risk reduction rely on the reduction of risk within the organisation by the implementatin of a loss control programme, whose basic aim is to protect the company's assets from wastage caused by accidental loss. Risk reduction strategies take two stages.

(a)   Collection of data on as many loss-producing incidents as possible provides information on which an effective programme of remedial action can be based. This process will involve the investigation, reporting and recording of incidents that result in death, injury, disease to individuals, damage to property, plant, equipment, materials or the product.

(b)   The collation of all areas where losses arise from the above incidents and the formulation of future strategies with the aim of reducing loss.

## *Captive insurance companies*

Increasingly large companies and organisations have felt the need to participate in their own insurance programme. One means of achieving this is the formation of a captive insurance company. Prior to such a decision being made, it is vital that a feasibility study be undertaken to establish the viability of such a proposal. A feasibility study entails an in-depth examination of the current insurance programme and loss experience with on-site surveys of all locations, together with identifying any capitalisation costs, taxation aspects, risk engineering services and reinsurance facilities necessary.

## *Risk management and safety*

There is no doubt that occupational health and safety is a cost item for any organisation and to every line manager in his own particular part of it. One of a manager's responsibilities is to balance each cost with a recovery. The cost of safety is measurable in accurate accountancy terms, such as:

(a) capital costs, including machinery safeguarding (guards, safety devices, etc.), fire protection (sprinkler systems, appliances, etc.)

(b) maintenance costs, in terms of keeping structures and plant to an acceptable level

(c) personnel costs, such as the training of full-time and part-time safety specialists

(d) time devoted by management, staff and others on matters of health and safety, e.g. the operation of health and safety committees

(e) information, instruction and training costs, such as induction training, fire drills, documentation of procedures.

Although it is difficult to balance these costs against specific return of income, management should be convinced that an investment in health and safety reduces the risk of these costs arising. Therefore, to promote health and safety within an economic budget requires risk management, namely the determination of priorities and spending wisely where it is needed to meet these priorities.

## THE MANAGEMENT OVERSIGHT AND RISK TREE (MORT) SYSTEM

During the period 1973–83 the United States Department of Energy developed and refined the MORT system safety programme. MORT is defined as 'a systemic approach to the management of risks in an organisation'. MORT incorporates methods aimed at increasing reliability, assessing the risks, controlling losses and allocating resources effectively.

### *The philosophy of MORT*

MORT philosophy is summarised in the following points.

#### Management takes risks of many kinds
Specifically, these risks are classified in the areas of:

(a) product quantity and quality

(b) cost

(c) schedule

(d) environment, health and safety.

**Risks in one area affect operations in other areas**
Management's job may be viewed as one of balancing risks. For instance, to focus only on safety and environmental issues would increase the risk of losses from deficiencies, schedule delays and costs.

**Risks should be made explicit where practicable**
Since management must take risks, it should know the potential consequences of those risks.

**Risk management tools should be flexible enough to suit a variety of diverse situations**
While some analytical tools are needed for complex situations, other situations require simpler and quicker approaches. The MORT system is designed to be applied to all of an organisation's risk management concerns, from simple to complex.

## The MORT process

The acronym, MORT, carries two primary meanings:

(a)  the MORT 'tree' or logic diagram, which organises risk, loss and safety programme elements and is used as a master worksheet for accident investigations and programme evaluations; and

(b)  the total safety programme, seen as a sub-system to the major management system of an organisation.

The MORT process includes four main analytical tools:

**Change analysis**
This is based on the Kepner–Tregoe method of rational decision-making. Change analysis compares a problem-free situation with a problem (accident) situation in order to isolate causes and effects of change. Change analysis is especially useful when the decision-maker needs a quick analysis, when the cause is obscure, and when well-behaved personnel behave differently from past situations, e.g. the Three Mile Island incident.

**Energy trace and barrier analysis (ETBA)**
ETBA is based on the idea that energy is necessary to do work, that energy must be controlled, and that uncontrolled energy flows in the absence of adequate barriers can cause accidents. The simple energy-barriers-targets concept is expanded with the details of specific situations to answer the question 'What happened?' in an accident. ETBA may be performed very quickly or applied meticulously as time permits.

**MORT Tree Analysis**
The third and most complex tool is MORT Tree Analysis. Combining prin-

ciples from the fields of management and safety, and using fault tree methodology, the MORT tree aims at helping the investigator discover what happened and why. It organises over 1500 basic events (causes) leading to 98 generic events (problems). Both specific control factors and management system factors are analysed for their contributions to the accident. People, procedures and hardware are considered separately, and then together, as key system safety elements.

### Positive (success) tree design

This technique reverses the logic of fault tree analysis. In positive tree design, a system for successful operation is comprehensively and logically laid out. The positive tree is an excellent planning and assessment tool because it shows all that must be performed and the proper sequencing of events needed to accomplish an objective.

## Objectives of the MORT technique

Fundamentally, MORT is an analytical procedure to determine the potential for downgrading incidents in situations. It places special emphasis on the part that management oversight plays in allowing untoward or adverse events to occur. The MORT system is designed to:

(a)  result in a reduction of oversights, whether by omission or commission, that could lead to downgrading incidents if they are not corrected

(b)  determine the order of risks and refer them to the proper organisational level for corrective action

(c)  ensure best allocation and use of resources to organise efforts to prevent or reduce the number and severity of adverse incidents.

The main areas of concern identified by the MORT Logic Tree are shown in Fig. 7.2.

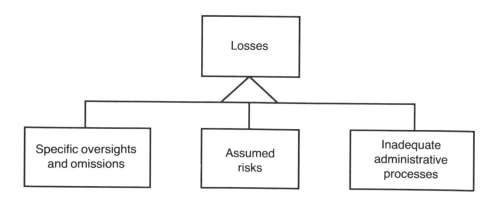

● **FIG 7.2  Areas of concern identified by the MORT Logic Tree**

## *The uses of MORT*

Because MORT was designed to fit together several parts of an organisation's efforts into a 'big picture', it can be used at all points of a system's life cycle. Beginning with the design of a project, for example, MORT stresses the importance of integrating the primary system components, namely people, procedures and hardware. Too often a facility is designed and construction is begun before procedures and people are considered. MORT's systematic approach to system configuration requires simultaneous consideration and development of people, procedures and hardware. In addition, the effects of changes in one of these primary components on other components are detailed through the use of the MORT method. Thus a review of design and design changes is a practical application of MORT.

The methodology from the MORT system is useful in project planning. The analysis trees are easily converted to project planning matrices and work flow charts, which are useful in determining the status of different parts of a system being made ready for operation. One of the most important uses of MORT is in emergency planning, a key way to reduce the severity of losses that do occur. The idea of 'operational readiness' may also be used as a current check on system operability. Quality assurance/quality control checks, safety assessments and trouble-shooting are all activities to which MORT contributes.

MORT is also instructive in how to monitor a system by illustrating the 'how', 'what' and 'why' of information collection, reduction and distribution. For instance, it shows how to evaluate the different methods of gathering reliability information and how to condense information to meet the needs of management. The risk assessment methods employed by an organisation and the means of independent review are also addressed by MORT, which shows how to balance the costs of redundancy against the benefits of independent judgement. It is useful, too, for planning the decommissioning of a facility or operation and component phase-out.

MORT can be used in all major accident and incident investigations, not only to discover the important immediate factors surrounding an accident ('what happened') but also to describe the management control factors ('why it happened') involved. MORT helps investigators to isolate quickly key accident contributors.

## TECHNIQUE FOR HUMAN ERROR RATE PROBABILITY (THERP)

THERP is a technique for predicting the potential for human error in an activity. It evaluates quantitatively the contribution of the human error component in the development of an untoward incident. Special emphasis is placed on the human component in product degradation.

The technique uses human behaviour as the basic unit of evaluation. It involves the concept of a basic error rate that is relatively consistent between tasks requiring similar human performance elements in different situations. Basic error rates are assessed in terms of contributions to specific systems failures.

The methodology of THERP entails:

(a) selecting the system failure

(b) identifying all behaviour elements

(c) estimating the probability of human error

(d) computing the probabilities as to which specific human error will produce the system failure.

Following classification of probable errors, specific corrective actions are introduced to reduce the likelihood of error.

The major weakness in the use of the THERP technique, however, is the lack of sufficient error rate data.

## RISK MANAGEMENT SERVICES

Risk management has a direct involvement with the insurance world and, indeed, many risk management companies operate as subsidiaries of the major insurance companies. As such, risk management companies provide a wide range of consultancy services which may vary from assisting in the solution of particular health and safety related problems to the preparation and implementation of risk management programmes.

Whilst the provision of advice to organisations in the fields of risk retention and risk transfer features as a significant part of the work of risk management companies, many of these companies provide consultancy services in the area of risk reduction. Risk reduction strategies can cover:

### Injury prevention
This area covers a wide range of issues, including the giving of advice on documentation requirements, such as those required for Statements of Health and Safety Policy and risk assessment, joint consultation procedures, safety monitoring systems, regulating contractors and their activities, health and safety training and the development of safe systems of work.

### Damage control
The development of hazard and damage reporting systems, together with systems for quantifying the cost of damage-only accidents.

### Security
The need for high standards of physical, technical and financial security has increased dramatically in the last decade. Typical losses can be associated with poor physical security of premises, pilfering of products, cash losses

during transfer, fraud and vandalism. The loss of confidential data can also represent a threat to an organisation.

### Vehicles
Motor fleet risk management involves the consideration of matters such as vehicle acquisition, driver selection, training and supervision, vehicle accident reporting, recording, investigation and analysis of claims, vehicle security and the operation and maintenance of vehicle fleets.

### Fire
Fire is the risk most common to businesses. In this case there is a need to consider fire prevention and control arrangements, including fire-fighting procedures and training, the handling and storage of flammable substances, basic requirements under the current fire safety legislation, such as means of escape, and emergency procedures.

### Product liability
Product liability is a branch of law concerned with the duties of those who manufacture, design, install, import, distribute and sell products. There are both general and specific requirements at common law and under health and safety, consumer protection and food safety law. Factors such as product design, utilisation, information and quality control are important areas of risk management in this field.

### Computers
Here consideration must be given to the physical protection of both hardware and software from fire and security risks.

### Public liability
The interaction between the company and third parties, such as members of the public, visitors and contractors of all types is an important area of risk management.

### Pollution
The potential for pollution of the air, water and ground, together with the risk of noise pollution, will, to some extent, depend upon the type of business operation. However, as society becomes more pollution-conscious, the duties under the current environmental protection legislation need careful consideration. Pollution incidents are bad for an organisation's public image.

### Occupational health
The last decade has seen considerable legislation directed both at reducing the health risks from work activities and at improving the health of employees and others who may be affected by the organisation's operations. The prevention of occupational diseases and conditions through control over hazardous substances, asbestos, noise and vibration, environmental and

biological stressors, radiation and other health risks must be considered as part of an overall risk reduction strategy.

### Business interruption

Any potential incident which could interrupt the business operation represents some form of risk. Typical examples of business interruption situations are machinery and plant breakdowns, production stoppages, the unavailability of spare parts for plant or raw materials, the provision of incorrect information by suppliers of goods and computer failures. For some organisations, this is a significant area of risk which must be adequately managed through planned maintenance programmes, quality improvement programmes and clear identification of individual responisbilities.

## CONCLUSION

Risk management embraces all those areas of risk to which an organisation can be exposed. The principal aim of a risk management programme is to provide a cost-effective system to protect the resources of an organisation by managing and controlling the risks that it faces. Many risks, currently covered through insurance, could well be taken on by companies using the range of risk management techniques available, thereby making substantial savings on insurance costs.

# Human Factors

The MHSWR, for the first time in the history of health and safety legislation, brings a human factors-related approach to occupational health and safety. While Reg. 4 places a general duty on employers actually to manage their health and safety activities through the effective planning, organisation, control, monitoring and review of the preventive and protective measures, Reg. 11 deals with the subject of 'capabilities and training'. The legal requirement relating to 'capabilities' is stated in Reg. 11(2) thus:

> Every employer shall, in entrusting tasks to his employees, take into account their capabilities as regards health and safety.

But what is meant by a 'capable' person? Various terms are found in the average dictionary: 'able', 'competent', 'gifted' and 'having the capacity'. Perhaps the last term is the most significant from a health and safety viewpoint. 'Capacity' implies both mental and physical capacity, mental capacity to understand why a task should be undertaken in a particular way, and physical capacity, the actual physical strength and fitness to undertake the task in question. Compliance with Reg. 11 requires an understanding, therefore, of human factors.

## WHAT ARE 'HUMAN FACTORS'?

The HSE's publication 'Human factors in industrial safety' (HS(G)48) defines 'human factors' as 'a range of issues including the perceptual, physical and mental capabilities of people and the interactions of individuals with their jobs and working environments, the influence of equipment and system design on human performance and, above all, the organisational charcteristics which influence safety-related behaviour at work'.

Although most British health and safety legislation places the duty of compliance firmly on the employer or body corporate, this duty can only be discharged by the effective actions of its managers. For instance, it is management's job to report and investigate accidents at work, but how frequently the causes are written down to 'operator carelessness', 'not looking where he was going' or, quite simply, 'human error', indicating that nothing can be done and no further action should be taken. Studies by the HSE's Accident Prevention Advisory Unit have shown that the vast majority of fatal accidents, and those causing major, injury could have been prevented by management action

For example:

1. During the period 1981–1985, 79 people were killed in the construction industry. Ninety per cent of these deaths could have been prevented. In 70% of cases, positive action by management could have saved lives.
2. A study of 326 fatal accidents during maintenance activities occurring between 1980 and 1982 showed that in 70% of cases, positive action by management could have saved lives.
3. A study of maintenance accidents in the chemical industry between 1982 and 1985 demonstrated that 75% were the result of management failing to take reasonable precautions.

These studies tend to emphasise the crucial role of the organisation in the management of job and personal factors, principally aimed at preventing human error.

There are three areas of influence on people at work, namely the organisation, the job and personal factors. These areas are directly affected by the system of communication within the organisation, and by the training systems and procedures in operation, all of which are directed at preventing human error.

## *The organisation*

Those organisational characteristics which influence safety-related behaviour include:

(a) promoting a positive climate in which health and safety is seen by both management and employees as being fundamental to the organisation's day to day operations, i.e. creating a positive safety culture

(b) ensuring that policies and systems which are devised for the control of risk from the organisation's operations take proper account of human capabilities and fallibilities

(c) commitment to the achievement of progressively higher standards, which is shown at the top of the organisation and through its successive levels

(d) demonstration by senior management of their active involvement, thereby stimulating managers throughout the organisation into action

(e) leadership, whereby an environment is created that encourages safe behaviour.

## *The job*

Successful management of human factors and the control of risk involves the development of systems of work designed to take account of human capabilities and fallibilities. Using techniques like job safety analysis, jobs should be designed in accordance with ergonomic principles so as to take into account limitations in human performance. Major considerations in job design include:

(a) identification and comprehensive analysis of critical tasks expected of individuals and appraisal of likely errors

(b) evaluation of required operator decision-making and the optimum balance between the human and automatic contributions to safety actions

(c) application of ergonomic principles to the design of man–machine interfaces, including displays of plant and process information, control devices and panel layouts

(d) design and presentation of procedures and operating instructions

(e) organisation and control of the working environment, including workspace, access for maintenance, noise, lighting and thermal conditions

(f) provision of correct tools and equipment

(g) scheduling of work patterns, including shift organisation, control of fatigue and stress, and arrangements for emergency operations

(h) efficient communications, both immediate and over periods of time.

## *Personal factors*

This aspect is concerned with how personal factors such as attitude, motivation, training, human error and the perceptual, physical and mental capabilities of people can interact with health and safety issues.

Attitudes are directly connected with an individual's self-image, the influence of groups and the need to comply with group norms or standards and, to some extent, opinions, including superstitions, like 'all accidents are Acts of God'. Changing attitudes is difficult. They may be formed through past experience, the level of intelligence of the individual, specific motivation, financial gain, and skills available to an individual. There is no doubt that management example is the strongest of all motivators to bring about attitude change.

Important factors in motivating people to work safely include joint consultation in planning the work organisation, the use of working parties or committees to define objectives, attitudes currently held, the system for communication within the organisation and the quality of leadership at all levels. Financially-related motivation schemes, such as safety bonuses, do not necessarily change attitudes, people frequently reverting to normal behaviour when the bonus scheme finishes.

Limitations in human capacity to perceive, to attend to, remember, process and act on information are all relevant in the context of human error. Typical human errors are associated with lapses of attention, mistaken actions, mis-perceptions, mistaken priorities and, in a limited number of cases, wilfulness.

### Safety incentive schemes

These are a form of planned motivation, the main objectives being that of providing motivation by identifying the targets which can be rewarded if achieved, and by making the rewards meaningful and desirable to the people involved. Any planned motivation scheme should always be viewed with care, however, in that it may alter behaviour, in order to win the rewards, but not necessarily attitudes. Safety incentive schemes are most effective from a health and safety viewpoint where:

(a) people are restricted to one area of activity, e.g. work in a production process

(b) measurement of safety performance is relatively simple

(c) there is regular rejuvenation or stimulation

(d) support is provided by both management and trade unions

(e) the scheme is assisted and promoted through safety propaganda and training

Important considerations to be taken into account prior to the introduction of safety incentive schemes are:

(a) they should be linked to some form of safety monitoring, e.g. safety inspections, safety sampling exercises

(b) realistic, measurable and achievable targets should be set

(c) on no account should incentive schemes be linked with accident rates, as this can discourage the reporting of accidents

(d) they tend to be short-lived and can get out of hand if not properly organised and monitored

(e) they can shift responsibility for health and safety from management to employees and others.

Much of the answer lies in effective communication on health and safety issues. Many staff see health and safety training as dull, uninteresting or unrelated to their specific tasks. In certain cases they do not understand why certain precautions are enforced by management or health and safety practitioners. Furthermore, many health and safety practitioners lack training in communication, seeing the running of health and safety training programmes as a chore. They may also not have the time to undertake training due to the demands of their full-time job.

As with many other areas of performance in organisations, there must be communication both vertically and laterally. The Board should set the objectives and standards which are both meaningful and measurable. They must communicate these objectives and standards down through the organisation and ensure such feedback of information as will enable them to measure and compare performance.

# Joint Consultation

Joint consultation between employers and employees is an important feature of the safety management process. Managers may liaise with trade union appointed safety representatives through the operation of a health and safety committee, or as part of a normal employer/employee consultative process.

## THE ROBENS REPORT 1970–72

The Robens Committee recognised the following:

1. The need for consultation born out of the principle of 'self-regulation'.
2. A lack of organised consultation on health and safety in many industries.
3. The constraints of traditional management/trade union roles where consultation did exist.
4. Out of the above constraints, the need to develop a more integrated management/workforce approach to health and safety at work.

This led to the inclusion of a framework for consultation in health and safety at work, namely the provision of Regulations covering the appointment, role and function of safety representatives and procedures for the establishment and functioning of safety committees, out of which came the Safety Representatives and Safety Committees Regulations 1977 (SRSCR). These broadly require an employer to consult and co-operate with safety representatives, and for safety representatives to co-operate with employers in connection with:

(a)  the arrangements for health and safety at work

(b)  measures for ensuring satisfactory health and safety performance

(c)  systems for checking the effectiveness of (a) and (b) above.

The Regulations are designed to provide a framework within which a business or undertaking can develop effective working relationships, which must cover a wide range of situations and activities.

The legal situation relating to joint consultation is covered by:

(a) *Health and Safety at Work Act 1974* – the law in principle

(b) *the Safety Representatives and Safety Committees Regulations 1977* – the detailed provisions of the law

(c) *the Approved Code of Practice* – the authoritative interpretation of the law

(d) *H.S.E. Guidance Notes* – guidance on the practical implementation of the law which have little or no legal status.

## SAFETY REPRESENTATIVES AND SAFETY COMMITTEES REGULATIONS 1977

The Regulations cover the following aspects:

### Appointment of safety representatives

A recognised trade union may appoint safety representatives from among the employees in all cases where one or more members are employed by an employer by whom the union is recognised. The employer must be notified by the trade union of the names of the safety representatives.

### Functions of safety representatives

Safety representatives have the following functions:

(a) to represent employees in consultation with employers

(b) to co-operate effectively in promoting and developing health and safety measures

(c) to make representations to the employer on any general or specific matter affecting the health and safety of their members

(d) to make representations to the employer on general matters affecting the health and safety of other persons employed at the workplace

(e) to carry out certain inspections

(f) to represent members in consultation with officers of the enforcement agencies

(g) to receive information from Inspectors

(h) to attend meetings of the safety committee if appropriate

It should be noted that none of these functions imposes a duty on safety representatives.

### Time off with pay

Employers must give safety representatives time off with pay to perform their functions and for any reasonable training they undergo.

### Inspections of the workplace

Safety representatives are entitled to carry out workplace inspections. Employers must give reasonable assistance in this case and notice in writing must be given by the safety representatives. The standard inspection of the workplace is on the basis of once every three months, although special inspections may be undertaken to inspect the scene of a reportable accident or dangerous occurrence.

### Inspection of documents

Safety representatives may inspect any document which the employer has to maintain, other than documents relating to the health records of identifiable individuals.

## *Approved Code of Practice*

The ACOP accompanying the Regulations covers the following issues:

### Qualifications of safety representatives

So far as is reasonably practicable, safety representatives should have had two years' experience with the employer or in similar employment.

### Functions of safety representatives

Trade union appointed safety representatives have a number of functions, namely:

(a)    to keep themselves informed of legal requirements

(b)    to encourage co-operation

(c)    to undertake health and safety inspections of the workplace and to inform the employer of the outcome of inspections

### Obligations of employers

Employers have an obligation to provide information to safety representatives on:

(a)    the plans and performance of the organisation with respect to health and safety at work

(b)    details of any hazards and the precautions necessary on the part of their members

(c)    the occurrence of accidents, dangerous occurrences and occupational disease

(d)    any other information, including the results of any measurements taken, e.g. as a result of air monitoring.

## SAFETY COMMITTEES

Safety committees can be an excellent form of joint consultation provided they are well-organised. Management commitment to ensuring that committee decisions are implemented, however, is essential if the committee is to have credibility with the workforce and achieve its objectives.

Safety committees have two principal objectives, namely to promote co-operation, and to act as a focus for employee participation.

### Functions of a safety committee

A safety committee should:

(a)  consider the circumstances of individual accidents and cases of reportable diseases

(b)  consider accident statistics and trends

(c)  examine safety audit reports

(d)  consider reports and information from the enforcement agencies

(e)  assist in the development of safety rules and systems

(f)  conduct periodic inspections

(g)  monitor the effectiveness of health and safety training, communications and publicity

(h)  provide a link with the Inspectorate.

A committee should be reasonably compact but should allow for representation of management and all employees. Management representation commonly includes line managers, supervisors, engineers, personnel specialists, medical and safety advisers. The safety committee should have authority to take action, and specialist knowledge should be available to assist the committee to make decisions where necessary.

It should be appreciated that a safety representative is not appointed by the safety committee nor vice versa. Neither is responsible to or for the other.

The following matters should be considered in planning the operation of a workplace safety committee:

(a)  the division of activities

(b)  the specification of clear objectives / terms of reference

(c)  clear definition in writing of the membership and structure

(d)  arrangements for publication of matters notified by safety representatives.

### Model constitution for a safety committee

As with any committee, it is essential that the constitution be in written form. A typical one is outlined on the next page.

# HEALTH AND SAFETY COMMITTEE
# CONSTITUTION

## 1    Objectives
To monitor and review the general working arrangements for health and safety, including the company Statement of Health and Safety Policy.

To act as a focus for joint participation in the prevention of accidents, incidents and occupational ill-health.


## 2    Composition
The composition of the committee will be determined by local management, but will normally include equal representation of management and employees, ensuring all functional groups are represented. Other persons may be co-opted to attend specific meetings, e.g. health and safety adviser, company engineer.


## 3    Election of Committee Members
The following officers shall be elected for a period of one year:

chairman

deputy chairman

secretary

Nominations for these posts shall be submitted by a committee member to the secretary for inclusion in the agenda of the final meeting in each yearly period. Members elected to office may be re-nominated or re-elected to serve for further terms. Election shall be by ballot and shall take place at the last meeting in each yearly period.

## 4    Frequency of Meetings
Meetings shall be held on a quarterly basis or according to local needs. In exceptional circumstances, extraordinary meetings may be held by agreement of the chairman.


## 5    Agenda and Minutes
The agenda shall be circulated to all members at least one week before each committee meeting. The agenda shall include:–


## (a)    Apologies for absence
Members unable to attend a meeting shall notify the secretary and make arrangements for a deputy to attend on their behalf.

## (b)    Minutes of the previous meeting
Minutes of the meeting shall be circulated as widely as possible and without delay. All members of the committee, senior managers, supervisors and trade union representatives shall receive personal copies. Additional copies shall be posted on noticeboards.

### (c)   Matters arising

The minutes of each meeting shall incorporate an Action Column in which persons identified as having future action to take, as a result of the committee's decisions, are named. The named person shall submit a written report to the secretary, which shall be read out at the meeting and included in the minutes.

### (d)   New items

Items for inclusion in the agenda shall be submitted to the secretary in writing, at least seven days before the meeting. The person requesting the item for inclusion in the agenda shall state in writing what action has already been taken through the normal channels of communication. The chairman will not normally accept items that have not been pursued through the normal channels of communication prior to submission to the secretary.

Items requested for inclusion after the publication of the agenda shall be dealt with, at the discretion of the chairman, as emergency items.

### (e)   Safety Adviser's Report

The Safety Adviser will submit a written report to the committee, copies of which shall be issued to each member at least two days prior to the meeting and attached to the minutes. The Safety Adviser's report may include, for example:

(i)    a description of all reportable injuries, diseases and dangerous occurrences that have occurred since the last meeting, together with details of remedial action taken

(ii)   details of any new health and safety legislation directly or indirectly affecting the organisation, together with details of any action that may be necessary

(ii)   information on the outcome of any safety monitoring activities undertaken during the month, e.g. safety inspections of specific areas

(iv)   any other matters which, in the opinion of the secretary and him, need a decision from the committee.

### (f)   Date, time and place of the next meeting.

# Developing a Safety Culture

A number of approaches to health and safety at work have been developed over the last forty years. There is the legalistic approach taken by many managers which fundamentally says, 'comply with the law, but no further than that!' This approach implies compliance with the legal standard solely to keep the organisation out of trouble. It implies the enforcement by management and safety specialists of strictly laid down procedures aimed at ensuring safe working. As such, it is doomed to failure as people at all levels have different perceptions and understanding of, and indeed respect for, legal requirements.

Secondly, there is the cost-benefit related approach, such as Total Loss Control, which says that all accidents, incidents and occupational ill-health represent substantial losses to the organisation. Under this system all loss-producing events, including property damage accident, which do not necessarily result in physical injury, are costed in terms of the direct and indirect costs. In certain organisations, particularly in the USA, such costs are written against an individual manager's budget as a means of motivating him to pay greater attention to health and safety in future. Such a system has merit in that it focuses attention on the need for improvement in the actual management of health and safety.

The third approach could be defined as the human factors approach. It is concerned with the various factors in the workplace which influence compliance with health and safety practices. Here we need to consider the influences created by the organisation, the actual jobs that people do and the personal or behavioural factors of individual operators. Considerable attention has been paid by industry recently to endeavouring to change the attitudes of workers with a view to improving performance, raising quality standards and increasing commitment to the success of the organisation. But how do we get these messages over to people engaged in the boring repetitive tasks, such as packing products or operating a particular machine, so commonly encountered in industry? Is the average operator interested in health and safety, or does he see it as yet another management imposition, for instance, making him wear hearing protection?

## WHAT IS A SAFETY CULTURE?

'Culture' is defined in several ways, for instance, 'a state of manners, taste and intellectual development at a time or place' and 'the refinement of mind, tastes, etc. by education and training'. Every organisation has its own particular culture, or set of cultures, which are developed over a period of many years. The statement 'This is the way we do things here' is largely an expression of an organisation's cultural approach to running the business. This principle must, therefore be, extended to the field of occupational health and safety if organisations are to manage their health and safety activities effectively.

### Establishing a safety culture

Clearly, there is more to the maintenance of good health and safety standards than mere legal compliance. Increasingly, the HSE is paying attention to the need to establish and develop a 'safety culture' within organisations. The HSE Director General's submission to the Piper Alpha Inquiry identified a number of important principles involved in the estblishment of a safety culture which is accepted by everyone and observed generally. Probably the most important principle is the acceptance of responsibility at and from the top, exercised through a clear chain of command, seen to be actual and felt through the organisation. This implies, firstly, clear definition of the roles, responsibilities and accountabilities of directors and senior managers for health and safety. Secondly, there must be conviction that high standards are achievable through proper management. Other factors raised by the Inquiry included:

(a)   the setting and monitoring of relevant objectives/targets, based on satisfactory internal information systems

(b)   the systematic identification and assessment of hazards and the devising and exercise of preventive systems which are subject to audit and review; in such approaches, particular attention must be given to the investigation of error

(c)   immediate rectification of deficiencies

(d)   promotion and reward of enthusiasm and good results.

### Developing a safety culture

The recent CBI publication 'Developing a Safety Culture' (1991) takes the above points further. This publication is based on a study undertaken by that organisation and identifies a number of features which are essential to a sound safety culture. A company wishing to improve its performance will need to judge its existing practices against them. The CBI stress that:

(a)  there must be leadership and commitment from the top which is gen-
uine and visible; this is the most important feature

(b)  there must be acceptance that improving health and safety performance
is a long-term strategy which requires sustained effort and interest

(c)  there must be a policy statement of high expectations which conveys a
sense of optimism about what is possible, supported by adequate codes
of practice and safety standards

(d)  health and safety should be treated as other corporate aims, and properly resourced

(e)  it must be a line management responsibility

(f)  'ownership' of health and safety must permeate at all levels of the work
force; this requires employee involvement, training and communication

(g)  realistic and achievable targets should be set and performance measured against them

(h)  incidents should be thoroughly investigated

(j)  consistency of behaviour against agreed standards should be achieved
by auditing, and good safety behaviour should be a condition of
employment

(k)  deficiencies revealed by an investigation or audit should be remedied
promptly

(l)  management should receive adequate and up-to-date information to be
able to assess performance.

This last point is important. How do organisations assess safety perform
ance? Is assessment based solely on accident incidence rates? Is there a company league table? Do managers of locations at the top of the league table get
reprimanded and those at the bottom their heads patted? Accident incidence
rate are only as good as the system for reporting accidents within the organisation. A recent HSC Annual Report (1990–1991) confirmed the long-estab-
lished fact that only approximately one-third of non-fatal injuries are being
reported to the enforcement agencies. On this basis, it is fairly certain that
many non-reportable injuries, which might be required to be reported under
the company's internal accident reporting scheme, are not being reported
either. In many cases, some managers are selective in terms of the reports
that they make, whereas others report all accidents.

Safety performance should be assessed on the basis of a formal audit or
review whereby improvement or deterioration in performance can be quantified in numerical terms. Such an audit or review should take place at least
twice a year and should be accompanied by an action programme identifying deficiencies and improvements required.

## The Du Pont approach to health and safety

The Du Pont organisation has long been recognised as one of the leaders in the field of occupational health and safety. Their approach is based on the 'Ten Principles of Safety' shown below.

1. All injuries and occupational diseases can be prevented.
2. Management is directly responsible for preventing injuries and illness, with each level accountable to the one above and responsible for the level below. The Chairman undertakes the role of Chief Safety Officer.
3. Safety is a condition of employment; each employee must assume responsibility for working safely. Safety is as important as production, quality and cost control.
4. Training is an essential element for safe workplaces. Safety awareness does not come naturally – management must teach, motivate and sustain employee safety knowledge to eliminate injuries.
5. Safety audits must be conducted. Management must audit performance in the workplace.
6. All deficiencies must be corrected promptly, through modifying facilities, changing procedures, better employee training or disciplining constructively and consistently. Follow-up audits must be used to verify effectiveness.
7. It is essential to investigate all unsafe practices and incidents with injury potential, as well as injuries.
8. Safety off the job is as important as safety on the job.
9. It's good business to prevent illnesses and injuries. They involve tremendous costs – direct and indirect. The highest cost is human suffering.
10. People are the most critical element in the success of a safety and health programme. Management responsibility must be complemented by employees' suggestions and their active involvement.

Management would do well to consider the above principles. At the end of the day the loyal, skilled and long-serving employee is the organisation's most valuable asset. Without these skills and loyalties many organisations would not be in business today.

## RAISING THE PROFILE OF HEALTH AND SAFETY AT WORK

One of the problems with health and safety is that it is seen by both management and staff as boring, uninteresting and not really their concern compared with, for instance, their main objectives which include staying in business, making a profit, increasing market share and reducing operating costs.

One of the ways of increasing awareness and raising interest is through the operation of a Health and Safety Award Scheme. Whilst there are a number of national award schemes run by safety organisations such as the Royal Society for the Prevention of Accidents (RoSPA) and the British Safety Council, many organisations run their own internal award schemes. An example of the criteria for a typical Health and Safety Award scheme follows:

## HEALTH AND SAFETY AWARD SCHEME

### Awards

Three Awards (Gold, Silver and Bronze) would be presented initially for the best health and safety performance by individual units during the year under review. After the first year of the Award Scheme coming into operation, an Award for Most Improved Performance would be added to the three Awards.

Each Award could take the form of a shield with badgelets, the shield passing from one winner to another on a yearly basis. The name of the winning unit each year would be engraved on a badgelet on the shield. The winner of the Gold Award could also be provided with a specially-designed flag to fly during the following year. Each Award would be accompanied by a framed certificate to be retained at the unit.

### Selection

Six finalists would be nominated by senior/area managers and a judging panel or group would visit each finalist unit with a view to assessing the winners and runners-up for the Awards.

### Recognition of Achievement

Maximum publicity would be given by the organisation to the Health and Safety Award Scheme through the use of posters and other means of raising awareness, and to the winners and finalists on a yearly basis. Some form of recognition for Award-winning units, by way of a special party, dinner, evening out for staff and part ners, or by giving Christmas hampers, should be made. Directors and senior management should show commitment to the Award Scheme by direct encouragement, identification personally with the Award Scheme and by attendance at Award-winners' functions.

### Lack of Commitment

Where there is clear evidence of lack of commitment to the Award Scheme by managers, which may be shown by a continuing deterioration in performance shown in the assessments, some form of stimulation, and even disciplinary action, may be necessary.

## *The significance of recognition of effort*

As with any field of management performance, recognition of effort by both groups and individuals is important. Such recognition of achievement can take a number of forms:

(a) publicity through the use of posters, in staff magazines and in the national press

(b) the presentation of certificates, shields, badges, trophies for display

(c) the provision of some form of celebration to which winners are invited

(d) attendance by directors and senior managers at award presenting functions to show their commitment to the award scheme.

## CONCLUSION

Developing and promoting a safety culture is an important aspect of health and safety management. It must be coupled with recognition of achievement, commitment from the very top of the organisation and a clearly defined system for assessing achievement which is both measureable and achievable by those concerned.

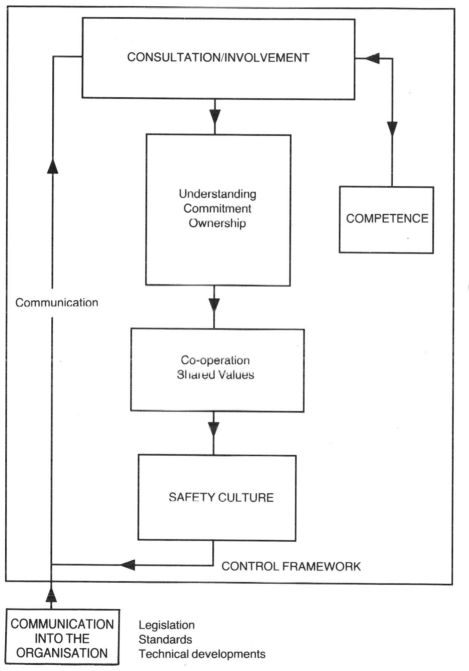

● **FIG 10.1 Promoting a positive health and safety culture**

# Bibliography

## Chapter 1

Health and Safety Executive (1989): Human Factors in Industrial Safety: HS(G)48: HMSO, London

Health and Safety Executive (1991): Successful Health and Safety Management: HS(G)65: HMSO, London

Health and Safety Executive (1992): Personal protective equipment at work: Guidance on the Personal Protective Equipment at Work Regulations 1992: HMSO, London

Health and Safety Guide no. 1 – Human Factors (London, 1994)

J.W. Stranks: *The Handbook of Health and Safety Practice* (London, 1994)

## Chapter 2

Secretary of State for Employment (1974): Health and Safety at Work etc. Act 1974: HMSO, London

Judge Ian Fife and E. A. Machin: *Redgrave's Health and Safety in Factories* (London, 1982)

M. Dewis and J.W. Stranks: *Health and Safety at Work Handbook* (Croydon, 1990)

Health and Safety Executive (1992): Management of Health and Safety at Work: Approved Code of Practice: Management of Health and Safety at Work Regulations 1992: HMSO, London

J.W. Stranks: *The Handbook of Health and Safety Practice* (London, 1994)

## Chapter 3

Health and Safety Executive (1989): Quantified Risk Assessment: Its Input to Decision Making: HMSO, London

Institution of Chemical Engineers (1989): Loss Prevention: Practical Risk Assessment: Student's workbook

Health and Safety Executive (1992): Display screen equipment work: Guidance on the Health and Safety (Display Screen Equipment) Regulatins 1992: HMSO, London

Health and Safety Executive (1992): Management of Health and Safety at Work: Approved Code of Practice: Mangement of Health and Safety at Work Regulations 1992: HMSO, London

Health and Safety Executive (1992): Manual handling: Guidance on the Manual Handling Operations Regulations 1992: HMSO, London

Health and Safety Executive (1992): Personal protective equipment at work: Guidance on the Personal Protective Equipment at Work Regulations 1992: HMSO, London

Health and Safety Executive (1992): Work equipment: Guidance on the Provision and Use of Work Equipment Regulations 1992: HMSO, London

## Chapter 4

F.E. Bird and G.E. Germain: *Damage Control* (American Management Association, 1966)

M.J. Crowe and H.M. Douglas: *Effective Loss Prevention* (Canada, 1976)

Chemical Industries Association Ltd (1977): A Guide to Hazard and Operability Studies

D.C. Petersen: *Techniques of Safety Management* (Kogakusha, 1978)

The Institution of Chemical Engineers (1985): Risk Analysis in the Process Industries

R.H. Amis and R.T. Booth: *Monitoring Health and Safety Management* (Institution of Occupational Safety and Health, 1991)

Health and Safety Executive (1992): Management of Health and Safety at Work. Approved Code of Practice: Management of Health and Safety at Work Regulations 1992: HMSO, London

R. Saunders: *The Safety Audit* (London, 1992)

J.W. Stranks: *The Handbook of Health and Safety Practice* (London, 1994)

## Chapter 5

Health and Safety Executive (1981): Health surveillance by routine procedures: Guidance Note MS18: HMSO, London

International Labour Organisation (1984): Occupational Health Services

J.P. Dees and R. Taylor: 'Health Care Management: a Tool for the Future', *AAOHN Journal*, xxxviii/2 (1990), pp. 52–8

Health and Safety Executive (1992): Management of Health and Safety at Work: Approved Code of Practice: Management of Health and Safety at Work Regulations 1992: HMSO, London

J.W. Stranks: *The Handbook of Health and Safety Practice* (London, 1994)

## Chapter 6

F.E. Bird and G.L. Germain: *Damage Control* (American Management Association, 1966)

J.A. Fletcher and H.M. Douglas: *Total Loss Control* (London, 1971)

F.E. Bird: *Management Guide to Loss Control* (Atlanta, 1974)

F.E. Bird and R.G. Loftus: *Loss Control Management* (Atlanta, 1976)

M.J. Crowe and H.M. Douglas: *Effective Loss Prevention* (Canada, 1976)

P. Morgan and N. Davies (1981): The Cost of Occupational Accidents and Diseases in Great Britain: Employment Gazette, HMSO, London

F.E. Bird and G.L. Germain: *Practical Loss Control Leadership* (International Loss Control Institute, 1986)

## Chapter 7

Health and Safety Information Bulletin No. 22: Risk Management: Prevention is Better than Cure (Industrial Relations Services, 1977)

Institution of Chemical Engineers (1985): Risk Analysis in the Process Industries

R.L. Carter, G.N. Crockford, and N.A. Docherty: *Handbook of Risk Management* (London, 1988)

Jardine Insurance Brokers Ltd: Risk Management – Practical Techniques to Minimise Exposure to Accidental Losses (London, 1988)

P. Madge: *The Risk of Pure Economic Loss Claims* (Risk & Insurance Management Review, 1988)

## Chapter 8

Health and Safety Executive (1989): Human Factors in Industrial Safety, HS(G)48: HMSO, London

Health and Safety Executive (1991): Successful Health and Safety Management, HS(G)65: HMSO, London

J.W. Stranks: *Health and Safety Guide No. 1: Human Factors* (London, 1994)

## Chapter 9

Secretary of State for Employment (1972): Report of the Committee on Safety and Health and Work, 1970–72 (Robens Report), Cmnd 5034: HMSO, London

Health and Safety Executive (1988): Safety Representatives and Safety Committees: HMSO, London

M. Dewis and J.W. Stranks: *Health and Safety at Work Handbook* (Croydon, 1990)

## Chapter 10

Confederation of British Industries (1991): Developing a Safety Culture: Business for Safety

J. Crunk (1991): Safety at Du Pont, a Cost Benefit Study (Du Pont De Nemours (Deutschland) Gmbh, Safety Management Services, Europe Postfach 1393, 4700, Hamm 1)

# Index